LANDSCAPE FIELDWORK

LANDSCAPE FIELDWORK

How Engaging the World Can Change Design

Gareth Doherty

University of Virginia Press
CHARLOTTEVILLE AND LONDON

The University of Virginia Press is situated on the traditional lands of the Monacan Nation, and the Commonwealth of Virginia was and is home to many other Indigenous people. We pay our respect to all of them, past and present. We also honor the enslaved African and African American people who built the University of Virginia, and we recognize their descendants. We commit to fostering voices from these communities through our publications and to deepening our collective understanding of their histories and contributions.

University of Virginia Press

Printed in the United States of America on acid-free paper

First published 2025

9 8 7 6 5 4 3 2 1

LIBRARY OF CONGRESS CATALOGING-IN-PUBLICATION DATA

Names: Doherty, Gareth, author.
Title: Landscape fieldwork : how engaging the world can change design / Gareth Doherty.
Description: Charlottesville : University of Virginia Press, 2025. | Includes bibliographical references and index.
Identifiers: LCCN 2024038386 (print) | LCCN 2024038387 (ebook) | ISBN 9780813952628 (hardcover) | ISBN 9780813952635 (paperback) | ISBN 9780813952642 (ebook)
Subjects: LCSH: Landscape architecture—Fieldwork.
Classification: LCC SB472 .D58 2025 (print) | LCC SB472 (ebook) | DDC 712—dc23/eng/20241118
LC record available at https://lccn.loc.gov/2024038386
LC ebook record available at https://lccn.loc.gov/2024038387

Cover art: aln2311/depositphotos.com

To Forster O. Ndubisi

If we opened people up, we'd find landscapes.

—Agnès Varda, *The Beaches of Agnès*

CONTENTS

ACKNOWLEDGMENTS

The advice of two colleagues at Harvard University's Graduate School of Design was instrumental in shaping this book. I thank Professor Niall G. Kirkwood for continually advising me on the importance of writing it and engaging in landscape fieldwork with me. Thanks to Professor John R. Stilgoe for inspiring me with his fieldwork on the so-called everyday landscapes. I also thank John for introducing me to Boyd Zenner, former senior acquiring editor at the University of Virginia Press. Thanks to Boyd and thanks too to Eric Brandt, director of the UVA Press. Sadly, Boyd passed away before I completed this book. I thank Boyd's successor, Mark Mones, for his careful, dedicated, patient, professional and thoughtful guidance.

I thank the Radcliffe Institute for Advanced Study at Harvard University for their support for the workshops that provided the discussions that inspired this book. I thank Lizabeth Cohen, former dean of the Radcliffe Institute, and Rebecca E. F. Wassarman, Radcliffe's executive director of academic ventures and engagement, for encouraging me to apply to host an exploratory workshop. When my application was rejected, I asked Paul D. Beran for feedback. The message was that my application was too well developed. I was encouraged to apply for an advanced-level workshop, which was subsequently successful and led to the development of the ideas in this book. Thanks to Sean Tath O'Donnell, who succeeded Paul as associate director of academic ventures for his support, and to Maura Madden and Wendy A. Frolich. I thank each of the participants for the workshops in April 2017 and May 2019, including Jonathan Shapiro Anjaria, Thomas Binder, Steven C. Caton, Galen Cranz, Dana Cuff, Dilip da Cunha, Namita Vijay Dharia, Wendy Gunn, Setha Low, George Marcus, Pol Fité Matamoros, Kathryn Moore, Belinda Tato, Jonas Tinius, Anne Whiston Sprin, Ed Wall, and Albena Yaneva.

Research for the book was facilitated by a Dean's Junior Faculty Research Grant at the Harvard University Graduate School of Design. I also acknowledge the assistance of the David Rockefeller Center for Latin American Studies, which made possible the field research in Brazil on Afro-Brazilian *terreiros* that constitutes the research for chapters 5 and 6; and the Center for African Studies for some of the research that informs chapter 6.

The publication was generously supported by a 2021 grant from the Graham Foundation for Advanced Studies in the Fine Arts. I thank Sarah Herda, director, James Pike, grant project manager, and Carolyn Kelly, grants manager, for their help and encouragement.

The book outlines projects in various parts of the world over a long timespan. As such, countless acknowledgments are due, and many of the instrumental informants are no longer with us. Many people helped me in the field with gracious and anonymous acts of kindness, and I will never be able to thank them. Thus, these acknowledgments are made in the spirit of incompleteness. I will list my main thanks according to chapter.

For the introduction, my sincere thanks go to Lady Suzy and Sir John Lewis. Lady Suzy recognized how moved I was by the landscape at Shute House and generously invited me to stay.

In Ireland, I thank Donegal County Council, Tony Brophy, Liam Campbell, Ann and Phil Coyne, Albert Doherty, Cathal Doherty SJ, Dessie and Pat Doherty, Vincent Lynn, David Lyons, Caoimhe and Sean McElligot, Tarla MacGabhann, Niall McGuinness, Marion and Paddy McGuinness, Con McLaughlin, Tony Monaghan, Eunan Quinn. I am grateful also to the late Sister Aloysius McVeigh, Sue Leopard, Joy and Liam McCormick, and Sir Geoffrey Jellicoe.

In the United Kingdom and the Netherlands, I thank Raoul Bunschoten and

Hélène Binet and their son Izaak and daughter Saskia, Tordis Berstrand, Roberto Bottazzi, Boris Brorman Jensen, Peter Hemmersam, Ioanna Marinescu, Mimi Sansone, and Corinna Till. Thanks also to Doris Sommer and Eduardo Davel for discussing aspects of the chapter with me.

In Bahrain, Mr. Abdullah and Mr. Kareem, Anny the Housekeeper, Emmanuel Lamort, and Camille Zakharia deserve special thanks. Chapter 3 is based on the afterlife of my previous book, *Paradoxes of Green: Landscapes of a City State,* and I thank Reed Malcom of the University of California Press for his editorial advice. Thanks also to Ahmed Dailami, Ali Akbar Bushiri, Nelida Fuccaro, and William Granara as well as Hamed Bukhamseen, Ali Karimi, Sh. Hamed Al-Khalifa, and the Sheldon Traveling Fellowship.

In The Bahamas, my thanks to Nicolette Bethel, Reg Smith, Eric Carey, Lynn Gape, Suneeta Rani Gill, Siri Linn Brandsøy, Paul Nakazawa, Padraic Kelly, Rob Daurio, Tomás Folch, Mariano Gomez Luque, Marcus Owens, Fadi Masoud, and Felipe Vera. Marisa Wells, photographer, who documented the project was always such a delight to work with. The project was generously facilitated by His Highness the Aga Khan.

In Bahia, I thank Marcela Franzen Rodrigues, Marcelo Moura Mello, Marina Correia, Sergio Ekerman, Adriano Mascarenhas, Duarte Vaz, and Elena Geppetti, and not least to André Lima O'Dwyer for translations. I enjoyed being in the field with Michelle Jean de Castro and Leonardo Finotti. Fábio Velame, director of the Faculty of Architecture and Urbanism of the Federal University of Bahia and director of the EtniCidades research initiative, kindly read and commented on my work. Thanks to the Afro-Latin American Research Institute, the Provost's Fund for Interfaculty Collaboration, the Center for the Study of World Religions at Harvard University. I acknowledge the intellectual and financial support of the Afro-Latin American Research Institute at Harvard University in the research for this chapter, and that of Harvard's David Rockefeller Center for Latin American Studies offices in Cambridge, Massachusetts, and São Paulo, Brazil. I also thank Helena Monteiro, executive director of Harvard University's David Rockefeller Center for Latin American Studies (DRCLAS) Brazil Office in São Paulo, and Tiago Genovese, program director in DRCLAS's Cambridge office. In particular, I wish to thank Professors Alejandro de la Fuente, Jacob K. Olupona, Kathleen Coleman, Bruno Carvalho, and

Doris Sommer. In Brazil, I thank Uiler Costa Santos, Caio Frederico e Silva, Thiago Montenegro Góes, Lucidio Avelino, and Vilma Patricia Silva. Thank you to Dulce Nascimento for so generously helping with the image of *Ficus insipida Willdenow* by her teacher, the late Maria Werneck de Castro. Oded Grajew and Mara Cardeal have been firm friends and supporters of my work, even joining me in fieldwork in Bahia. And most of all, I thank the community of the Casa de Oxóssi for welcoming me.

Thanks to many colleagues, friends, and mentors at Harvard Graduate School of Design, including Francesca Benedetto, Anita Berrizbeitia, Eve Blau, Danielle N. Choi, Diane E. Davis, Jill Desimini, Ed Eigen, Ann Forsyth, Stanislaus Fung, Gary R. Hilderbrand, Karen Janosky, Jerold Kayden, Alex Krieger, Ali Malkawi, Rahul Mehrotra, Mohsen Mostafavi, Paul W. Nakazawa, Chris Reed, Peter G. Rowe, John R. Stilgoe, Martha Schwartz, Charles Waldheim, and Sarah M. Whiting for stimulating and challenging conversations. Some of the material quoted was translated with the help of the Graduate School of Design's Racial Equity and Anti-Racism (REA) Fund.

Over the years, I have had the honor and pleasure of working with accomplished and inspiring artists and photographers, including David Lyons, Hélène Binet, Camille Zakharia, Leonardo Finotti, Marisa Wells, and Adolphus Opara, whose photographs illustrate the chapters.

I was with Shaheen Pirzada when I had the idea for the structure of the book. Shaheen is an astute landscape fieldworker who walks and talks regularly. Many others helped in so many ways, among them Anne and Sudip Chakraborty, Jock Herron, Julia Moore, Levi Herron, and Li Hou. Anna Lambertini invited me as a visiting professor at the University of Florence for a series of lectures which allowed me the opportunity to test some of the ideas presented in the book. Eleonora Giannini, Sergio Zorzetto, and Antonella Valentini, among many others were incredibly kind and generous during my stay in Florence. Sihem Lamine of the Harvard Center for Middle Eastern Studies' Tunis Office challenged me to have a subtitle.

Thanks to many colleagues, friends, and mentors at Harvard and MIT for their comments and conversation that have helped shape this book in so many ways. They include Daniel E. Agbiboa, Emmanuel K. Akyeampong, Ali Asani, Bruno Carvalho, Steven C. Caton, Sidney Chaloub, Kathleen M. Coleman, Alejandro de la Fuente, Henry Louis Gates Jr., William Granara, Hashim Sarkis, Tarek E. Masoud,

George Paul Meiu, Jacob K. Olupona, Nasser Rabbat, Doris Sommer, Anne Whiston Sprin, Ajantha Subramanian, and James Wescoat. I had the pleasure of doing fieldwork with Daniel E. Agbiboa, Emmanuel K. Akyeampong, Steven Caton, and Anne Whiston Sprin, and of meeting Professor Olupona in the field.

Thank you to those landscape architects, designers, and fieldworkers I interviewed while writing the book: they helped me realize that there was another project on landscape fieldworkers. They include Olatunji Adejumo, Stig L. Andersson, Sierra Bainbridge, Nelson Brissac, Philippe Coignet, Momoyo Kaijima, Jala Makhzoumi, Laurie Olin, Jungyoon Kim, Yoonjin Park, Mary Robinson, Jorge Silvetti, Ziying Tang, Duarte Vaz, Günther Vogt, Ed Wall, Diana Weisner, and Sara Zewde.

Some of the chapters in the book build upon a seminar, which then became a lecture course cross-listed between three departments: Landscape Architecture, African and African American Studies, and Anthropology. I thank the respective department chairs: Anita Berrizbeitia, Alejandro de la Fuente, and Ajantha Subramanian for supporting the course, and Dean Sarah Whiting for suggesting it. At the time of writing, that lecture course was integrated within the core Master in Landscape Architecture history and theory curriculum at the Harvard Graduate School of Design. I thank my department chair, Gary R. Hilderband, and my co-instructor, Charles Waldheim, for their collegiality. And thank you to the students in those courses for many great conversations.

I thank those master's thesis and doctoral students I have had the good fortune to work with for challenging me directly and indirectly on the contents of this book. They include Aziza Abdulfetah, Weaam Alabdullah, Rawan Alsaffar, Hamed Bukhamseen, Daniel Daou, Jake Delucca, Vanessa Harden, Junnan Mu, Pedro Rodriguez-Parets Maleras, Renugan Raidoo, Roberto Ruiz Ransom, Pablo Pérez-Ramos, Elaine Stokes, Sophia Xiao, and Ayaka Yamashita.

I thank the numerous research assistants and research associates who helped with aspects of this book. They include Rawan Alsaffar, William Baumgardner, Pol Fité Matamoros, Miguel López Meléndez, Suryani Oka Dewa Ayu, Rajji Desai Sanjay, Junho Kang, Isaiah Krieger, Fatma Mhmood, Forrest Rosenblum, Skyler Smith, Chenlu Wang, and Ayaka Yamashita. Thanks to Pol Fité Matamoros for his critical comments and provocations. Yonghui Chen, postdoctoral fellow, deserves a special thanks for his ingenuity, tenacity, and incredible attention to detail, which were of great help in the crucial last stages of the project. And I thank Hoon Kim and Haejung Choi of Why Not Smile LLC, New York, for many years of collaborations on graphic design and communication.

Special thanks go to Emily Sekine for her astute and careful editing and for being such a delight and pleasure to work with. And thank you to Milton Kornfeld also for his critical eye on for parts of the text. Maura High copyedited the final version with precision and professionalism. Valerie J. Cummings tried to keep me on schedule. And thank you to the late Henry Grant SJ for his encouragement to study the ecology of people and places, and landscapes' transcendent qualities.

The draft manuscript was reviewed and debated during a book manuscript workshop in February 2022 facilitated by a grant from the Weatherhead Center for International Affairs (WCFIA) at Harvard University. I thank the readers for their astute comments, advice, and critique and for their conversation over two days that helped shape the final manuscript. They are Andrea Ballestero, Jala Makhzoumi, Marc Treib, Ed Wall, and Albena Yaneva. Mark Mones facilitated the discussion. Ed Wall, who joined us from his vacation, described it as "the best two days of a weekend/holiday that I have spent in ages." I thank Melani Cammett, the director, and Ted Gilman, former executive director, and Sarah Banse and Michelle Nicholasen of the WCFIA for their support.

In addition to the readers in the book manuscript workshop, I thank the anonymous readers who reviewed the initial book proposal, and then the full manuscript, for their critique and helpful and insightful advice.

Thanks to those colleagues, friends, and mentors who freely gave of their time to discuss aspects of this book throughout its development. They include Frederick "Fritz" Steiner, Tao DuFour, Steven N. Handel, John Beardsley, Susan Nigra Snyder and George E. Thomas, Kenneth Frampton, Anne Whiston Sprin, Kathryn Moore, and the late Forster O. Ndubisi. Forster insisted we talk every three weeks until COVID-19 tragically intervened and took Forster away from us much too soon. Because these conversations with Forster were so formative in shaping this book, I dedicate the book to Forster's memory.

Moisés Lino e Silva has been the most constant collaborator and co–landscape fieldworker, sharing many of the fieldwork experiences in this book. Ivone Pereira de Andrade also joined us for part of the landscape fieldwork.

THE I IS A DOOR

A Note on Use of the First Person

Landscape fieldwork sits between the descriptive capacity of ethnography and the prescriptive aspects of design. An essential distinction between ethnography and design is the point of view of the author. Ethnography accepts the centrality of the fieldworker, and as a result ethnographies are usually written in the first person. In design scholarship, by contrast, we often look for neutrality, separating feelings from objective facts, a position inherited from art history and allied fields.[1] I use first person throughout in part to illustrate how, as a landscape fieldworker, I become deeply immersed and invested in each field site and its users. While the positionality of the fieldworker matters to the narrative, the audience and subject for fieldwork are usually the inhabitants of the landscape. Most importantly, one intention of writing in the first person is to bring you, the reader, along with me into the various projects. The architectural ethnographer Albena Yaneva describes this form of ethnographic writing as akin to "a sketching technique that provides the basis for painting specific scenes, profiles, and events."[2] Ethnographic sketching allows the landscape fieldworker to include various layers of information that offer contextual grounding for a project. Tim Ingold, the anthropologist, describes the process of sketching as an accumulation of lines; when you sketch on top of a line, the first line is still visible; there's always room and potential for more.[3] If ethnographic writing is open to this generative potential, it has similar aims to sketching.[4] For ethnographers I may have too few of these details, for designers on the other hand, there may be too many.

Throughout the book, I also reflect upon what I've observed in order to learn from that knowledge and ask how it can be applied in the future.[5] The anthropologists John Monaghan and Peter Just point out in their excellent *Introduction to Social and Cultural Anthropology* that all scientists begin their experiments by calibrating their instruments; however, they point out that it is much easier to calibrate a spectroscope than a person.[6] In self-calibration, it is important to acknowledge the significance of potential biases in observing, describing, and interpreting data. Recognizing these biases has become part of the ethos of anthropology since the 1980s especially, with the publication of *Writing Culture* by James Clifford and George Marcus and other texts.[7] Ethnographers have been grappling with the colonial legacies of the field by turning away from a more positivist scientific perspective, which assumes objectivity is always possible, toward more reflexive modes of engagement. In this book, I also aim to highlight the challenges in the distinction between observer and observed, subject and object. I write in the first person because it is the ethical way to write about these landscapes. Subject and object are not separated, as on the top line below; rather they are linked together:

Subject | Object
Subject—Object

In discussing the relationship between the subject and object, the ecofeminist Donna Haraway writes: "Situated knowledges require that the object of knowledge be pictured as an actor and agent, not as a screen or a ground or a resource, never finally as slave to the master that closes off the dialectic in his unique agency and his authorship of 'objective' knowledge."[8] In acknowledging the deep experiential entanglements between people and the land they inhabit, Haraway recognizes that landscapes by their very nature are situated and cannot be adequately described in terms of slave to the master, or subject versus object.

I recognize and accept the fact that I cannot totally separate myself from the field, especially since I rely on my embodied engagement to perform landscape fieldwork.

The immersive experience, or "embodied engagement," of landscape fieldwork is best communicated through the "I." The landscape fieldworker cannot hide behind the veil of authority, using the "royal we" to depersonalize the work. Instead, the personal perspective provides a layer of accountability. As Kenny Cupers and his coauthors write in their book *What Is Critical Urbanism?,* "Placing oneself in the picture is not an act of solipsism, it is a way to move out of the center to make space for others."[9]

Immersive fieldwork narrated by the fieldworker opens possibilities for describing the "thickness" of a multirelational, tangible and intangible, landscape. For this reason, there is a long tradition within landscape architecture of writing in the first person that extends from the founding of the field through to the present today. Unsurprisingly, the first person is often used in field reports and surveys, but it is also used in more conventional writing and, increasingly, in academic texts. Writing in a range of contexts, examples include James Corner,[10] Lawrence Halprin,[11] Randolf T. Hester,[12] J. B. Jackson,[13] Gertrude Jekyll,[14] Geoffrey Jellicoe,[15] Ian L. McHarg,[16] Jala Makhzoumi,[17] Laurie Olin,[18] Frederick Law Olmsted,[19] Anne Whiston Spirn,[20] John R. Stilgoe,[21] Ed Wall,[22] Tim Waterman,[23] and Günther Vogt.[24] Anne Whiston Spirn sometimes writes her notes in the field as an embodied response to the landscape within which she is situated. Sometimes these notes are directly included in her publications,[25] for example, Spirn's description of the Forest Cemetery in Stockholm in *The Language of Landscape.*[26]

Spirn's eBook, *The Eye Is a Door,* uses photography as the lens through which to read and understand a landscape. Spirn writes, "To photograph mindfully is to look and think, to open a door between what can be seen directly and what is hidden and can only be imagined."[27] Referencing the poet Seamus Heaney's characterization that words are doors, and through her own words and photographs, Spirn brings the reader into the landscapes that she was immersed in. For Spirn, "the eye is a door." The use of "I" in this volume is like the "eye." You might say, "the I is a door."

So, this preface concludes with a disclaimer: I use first person because it's important epistemologically. This is not a book about me, the author, and I hope you will not interpret it that way. Instead, the book tells the story of various landscapes through the eyes of the landscape fieldworker, who happens to be me. The reader will see me in different forms: as a student interviewing Jellicoe; a designer working in my hometown; then working in an office in London; then as a doctoral student completing his dissertation; then as a full-time researcher and professor. My history is intertwined with these cases, and I acknowledge that entanglement rather than ignore it. There is a lesson here for fieldwork because it is nigh impossible to do any form of immersive fieldwork without projecting our own histories and values as human beings on other human and nonhuman agents. What we must do, though, is recognize that other people, and even nonhumans, also have agency. Through quiet fieldwork, we can begin to listen to that agency. As you accompany me in learning about these various places and communities, I ask that you keep in mind the centrality of the fieldworker to landscape fieldwork and reflect upon and acknowledge your positionality relative to the landscapes in question.

NOTE ON TRANSLITERATION AND TRANSLATION

This book includes references to several languages, and dialects, in which I have various degrees of competency. They include Irish, French, Latin, Arabic, Persian, Bahamian Creole, Brazilian Portuguese, Brazilian Yoruba, and Yoruba. In all cases, I have tried to maintain a generally standard spelling, with some modifications.

For translations from the Irish, I use de Bhaldraithe's English-Irish dictionary, https://www.teanglann.ie/en/eid/, which differs slightly from the official standard, the *Caighdeán Oifigiúil.*

Transliterations of Modern Standard Arabic and Bahraini dialect are modified from the conventions of the *International Journal of Middle Eastern Studies* (IJMES). Specifically, IJMES conventions were followed for technical terms and expressions and vocabulary in Arabic, but proper and place names conform with their general attestation in English spelling in a Bahrain-related context. For example, rather than use the IJMES spelling *shaykh,* I use *sheikh,* which is standard in the Arab states of the Persian Gulf. Many place and institutional names in Bahrain have generally "accepted" English spellings, which I use, that do not conform to a standard and have many varieties, for example, *Riffa,* rather than *al-Rifaʿ.* Personal names like *Alireza, Isa,* and *Latif* are generally accepted spellings of these names in English.

Afro-Brazilian terms in Portuguese and Brazilian Yoruba are modified from the book *Meu tempo é agora* (My time is now) by Mae Stella de Oxóssi (writing as Maria Stella de Azevedo Santos, Salvador da Bahia: Assembleia Legislativa do Estado da Bahia, 2010).

With the exception of a few public figures, all names in this book have been changed.

PROLOGUE

Border Crossings

Without fail, every Saturday, right after lunch, my mother would bring my brother and me to visit our grandmother across the border in Northern Ireland. Dishes washed and dried, we would hop into the family's red Ford Fiesta, and hurry up the road. My brother and I were always in the back seat.

First, we would drive through Glentogher, a narrow valley formed by a glacier. As we went south, the road would get progressively steeper. The colors of the fields would change, from the year-round green grasses in the valley to the beiges, browns, and purple heathers of the higher ground. At a certain point, at Kerrykeel Brae, we would cross to the other side of the hill. There, on the horizon in front of us, Northern Ireland would appear across the narrow blue estuary called Lough Foyle. As we reached the bottom of the hill, we would drive along the water's edge as we approached the border.

This was at the height of the Troubles, the armed conflict that dominated both sides of the Irish border for thirty years starting in the 1960s. The border had existed since the winter of 1921–22, when Ireland was partitioned into Northern Ireland—which remained part of the United Kingdom—and the Irish Free State, later to become the Republic of Ireland. Crossing the border from the Republic, where we

Figure 1. The road to Derry, Inishowen, Co. Donegal, Ireland. (Photograph by Niall McGuinness)

lived, meant going from the mainly Catholic Free State to the Protestant-controlled Northern Ireland, which had a large Catholic minority. At that time, Catholics in Northern Ireland were amid a struggle for civil rights and fair employment. Meanwhile, the Republic of Ireland asserted a territorial claim over Northern Ireland, a claim the British state found untenable. Even our destination had two names: Derry to Catholics and Londonderry to Protestants. (The Gaelic, *Doire,* was anglicized to Derry, and then prefixed with London as an honorific in the 1600s. Today, in an awkward compromise, the city is referred to as Derry-Londonderry.) My mother was born just a few yards across the border in a Northern Irish townland called Culmore, *Cúil Mór,* meaning the Big Woodland in the original Irish, though the woodland was long gone. Culmore had a mixed community, unlike most of Derry-Londonderry which was divided into Catholic and Protestant areas.

Right before the border, we would drive past the customs' checkpoint. The customs did not stop cars entering Northern Ireland. The UK economy was in general much bigger and stronger than the economy of Ireland, which in the 1970s had a population of just three million or so citizens. Across the border in the North, prices were lower and consumer choices greater. So, the customs officials did not entertain the possibility that anyone would want to smuggle goods *into* the North; smuggling was considered a one-way affair.

First thing after crossing the border, we always stopped at Harkin's petrol station where we would fill up the tank for the week ahead. Petrol, or gas, for the car was one commodity whose point of origin was difficult for the customs officers to decipher should they decide to do a spot check on the way home. For this reason, it was best not to keep receipts.

Though crossing the border was easy enough, we still had to pass through the British Army checkpoint, where the officers were very concerned with every car that passed through. That's where things got dicey. The British Army were there ostensibly to keep the peace between the two opposing sides in the Troubles, but for many on both sides of the border, their presence added to the conflict. As we approached the checkpoint, just a quarter of a mile away from my granny's, my mother would get tense and tell us to keep quiet and behave.

At the checkpoint the soldiers would verify documents and look at the trunk of the car. Occasionally, they would take vehicles aside for a more thorough searching. My mother would dread the car being searched, which happened often. Not alone was it stressful, it could set us back by an hour or more and delay the whole afternoon.

I was very conscious, from a young age, that there were two types of people: Catholics and Protestants. To an outsider we probably all look the same, but a northern Irish person learns to read the cues over names, accents, and haircuts. Before marrying my father and moving across the border, my mother had worked for the Labor Exchange in London. Being a Catholic (albeit with Protestant heritage), she was unable to find employment in the Northern Irish civil service. At the Labor Exchange on Horseferry Road, my mother was responsible for pairing employers with staff—a necessary bureaucracy in the postwar United Kingdom. She often told us how it was common in London at that time to receive job advertisements specifying "No Blacks, No Irish, Need Apply." When she received such a stipulation from a racist employer, she would pair the worst possible candidate with the job.

Invariably, by the time we would arrive at my grandmother's house, my mother was shaken and infuriated from the British Army checkpoint. After calming down over cups of tea, my mother would then drive us to my aunt's house a few miles away closer to the city center. There, we might meet my cousins, who rode motorcycles. We would then go shopping in Superfare, the supermarket, to stock up on British brands such as McVitie's Digestives, Mr Kipling cakes, Dale Farm ice cream, Ambrosia creamed rice, and Flora margarine—brands that I was to later discover permeate the postcolonial world. We would fill the trunk full of groceries, making sure to squeeze as many of our smuggled items as possible into the spare tire compartment so the customs would not see them on the way home.

After stocking up at Superfare, we would go to the city center and watch the riots. Derry-Londonderry is one of the last walled cities in Europe. The Foyle River divides the city in two: the Waterside and the Cityside. At that time, many of the curbstones in each sector were marked in the national colors of red, white, and blue for Protestants in the Waterside, and green, white, and orange for Catholics in the

Figure 2. Border checkpoint, Northern Ireland. (Photograph by Alain Le Garsmeur, "The Troubles" Archive / Alamy Stock Photo)

Cityside. At that point in time, it would have been unsafe for us to venture into the Waterside with a southern-registered vehicle.

The riots regularly took place in the Bogside, a stretch of land outside the walled city, near the site of Bloody Sunday, otherwise known as the Bogside Massacre, in which twenty-six individuals were killed in conjunction with a civil rights march on January 30, 1972. Immortalized in U2's famous song, Bloody Sunday was the moment from which time was measured for many residents, including my grandmother. This would have been seven or eight years after Bloody Sunday. My mother would park her car in this place steeped in the rebellion and the struggle for equal rights, and we would watch from a safe distance. Against a backdrop of the city walls on the hilltop, we'd see youths throwing stones at the army in their tanks. Sometimes petrol bombs would explode.

Most of my friends back at home had never even crossed the border and were shocked when I would report what I saw at school on Monday. And I did think it a bit strange. Eventually, I would ask my mother why she was making us watch these riots. She replied simply: "It's important you know what is happening on your own doorstep."

This experience has stayed with me all my life. Watching the riots did not necessarily make me more sympathetic to either the Irish Nationalist or British Unionist political causes. That wasn't the point. My mother's small and simple act of defiance of making her young sons watch the riots instilled in me a deep appreciation for the embodied experience of being in "the field." I still remember the smell of smoke. The vision of a drab rainy Saturday afternoon illuminated with a flashing petrol bomb. The sound of sirens and the shouts of the rioters.

Seeing the riots in person, rather than merely on the television screen like most of my classmates, helped to demystify them. By making the riots tangible, my understanding of what was happening could move beyond the violent spectacle to the deeper questions of national unity and struggles for power that lay beneath. I could begin to see the riots as part of a much more complicated series of relationships that were not necessarily the binary Irish versus British, Catholic versus Protestant, but deeply rooted in a common struggle for social justice and fairness.

Later in life, I came to understand how landscape was at the heart of this struggle. People's differing relationships to land was a central, yet often overlooked, driver

Figure 3. British Army controlling rioting on the streets of Derry-Londonderry. (Photograph: Alain Le Garsmeur "The Troubles" Archive / Alamy Stock Photo)

of the Northern Irish Troubles. To boil it down: When the British began invading Ireland in the seventeenth century, they challenged the collective farming practices of the indigenous Irish people. They introduced new agricultural practices and concepts of private land ownership. They claimed the Irish had no right to lands that they did not farm to their advantage. Mary Bagenal, a Protestant character in Brian Friel's play *Making History,* puts it this way: "You talk about 'pastoral farming'—what you really mean is neglect of the land. And a savage people who refuse to cultivate the land God gave us have no right to that land."[1] The collectively farmed pastoral lands were claimed by the British. In the Plantation of Ulster, the land was colonized, or "planted," with Protestants loyal to the Crown, disenfranchising the indigenous Irish, and thus ensuring centuries of strife. To this day, residents maintain that Protestant-owned lands are easily distinguished from their Catholic neighbors for their neatly trimmed hedgerows whereas Catholic-owned lands are more unruly and unkempt.

The space of the Bogside, where my family went to watch the riots, was mostly inhabited with descendants of the disenfranchised indigenous Irish who were lured to Derry-Londonderry by the linen and shirt-making industries of the nineteenth century. Seen from a spatial perspective, the riots were more than a disgruntled disenfranchised youth demonstrating their frustration toward the mighty British army. These events were the enactment of public space. "Space is a practiced place," declared Michel de Certeau, the French philosopher of everyday life.[2] The riots transformed the spaces of the Bogside into practiced places.

The Irish / British tensions were eventually resolved through negotiations that led to a ceasefire. Rethinking shared space and landscape was a central part of these negotiations. As part of the Peace Process—which began with a ceasefire by the IRA in 1994 and culminated in the Good Friday Agreement of 1998—the governments invested in postconflict placemaking strategies. These landscape interventions included the removal of symbols and barriers in politicized spaces like the Bogside, and the demilitarization of conflict architecture. They also encouraged urban regeneration and building up capacities for healing and reconciliation though mural reimaging (such as depicting sports figures instead of political figures), conflict tours, and community festivals and gardens.[3]

But some divisions were more challenging to address. One intractable aspect of the Northern Irish conflict was the Republic's claim over Northern Ireland. The Constitution of Ireland (the Republic) defined the national territory as consisting of "the whole island of Ireland, its islands and the territorial seas." This claim could not be dropped without disenfranchising 40 percent of the Northern Irish population who wanted to part of the Irish nation. Finally, in a national referendum in 1998, 94 percent of the Irish electorate voted to replace the basis for defining the Irish nation from territory to the people born in the territory. The constitution now reads: "It is the entitlement and birthright of every person born in the island of Ireland, which includes its islands and seas, to be part of the Irish Nation. That is also the entitlement of all persons otherwise qualified in accordance with law to be citizens of Ireland. Furthermore, the Irish nation cherishes its special affinity with people of Irish ancestry living abroad who share its cultural identity and heritage."[4] This solution that came out of the negotiations holds an important lesson for landscape architecture: it was only when the nation was reconceptualized to recognize *people* as an integral component of territory that the long-held divisions were broken. Viewing landscape as the living embodiment of people, a space for the practice of life, allows us the opportunity to operationalize landscape as a social, cultural, and political construct. When we accept this multidimensionality of landscape, we release landscape's potential to bring about a more just, beautiful, and equitable world.

There are countless examples from around the world of landscape architecture claiming its agency in the resolution of conflicts. In post-apartheid South Africa, for instance, Tarna Klitzner Landscape Architect (TKLA) designed a series of public spaces to foster community interaction and social and environmental equity after the end of apartheid. The project was supported by the Urban Strategists, Violence Prevention through Urban Upgrading (VPUU), and TKLA's work was embedded within the VPUU construct.[5] Similarly, the landscape architect Jala Makhzoumi considers how landscape's intangible qualities can bridge cultural divides and offer political empowerment within the context of postwar reconstruction in Iraq and Lebanon.[6] Makhzoumi's work highlights how in the Middle East's war-torn landscape, landscape architecture can both be regarded as an agent of Western colonialism and neoliberalism, and simultaneously deeply rooted in the national imaginary. Makhzoumi's work among the Kurds of northern Iraq seeks to work within their understanding of landscape, rather than imposing an external perception of landscape.[7] These are just two examples of landscape architecture as a practiced place that serve as inspiration for this book. That early fieldwork with my mother and brother helped me to understand the shaping of landscape as being fundamental to bringing about a fairer and more equitable world.

Back in the day, in the Irish Northwest, we would make our way home on Saturday evenings, traveling north to go south. The sequence was reversed: the British Army checkpoint was not interested in our departure, but the Irish Customs would check the car for smuggled goods. I always wondered if they knew about the groceries packed into the spare tire compartment and just didn't say anything. A mother and two young sons—smugglers and passive protestors—went home to the relative quietness of village life, carrying not just Mr Kipling and McVities but also vivid memories of the practice of landscape.

INTRODUCTION

A Landscape of Fieldwork

Get out now. Not just outside, but beyond the trap of the programmed electronic age.

—John R. Stilgoe

Landscape architecture is at a crossroads. On the one hand, the field's ability to combine diverse, interdisciplinary insights and perspectives from architecture, design, environmental studies, the social sciences, and other fields puts it in an excellent position to address some of the most pressing global issues of our time, including climate change and social inequities. But the field's increasing reliance on digital and technological solutions to design challenges means that landscape architects are often not taking the time to fully understand the ways that humans use spaces. In addition, while landscape is architected all over the world, the discipline of landscape architecture remains centered on North America, Europe, and parts of Australasia. This book aims to change that by, through real-life examples, offering tools and practices for landscape architects to engage more deeply with multidimensional, diverse landscapes and the communities that create, live in, and use them.

Echoing John R. Stilgoe, the historian of the everyday, the book encourages readers to go out and do their own fieldwork in the sites where they are working.[1] Landscape fieldwork—as a form of measurable, immersive, multimedia, and imaginative interactions with a site—is and should be at the core of what landscape architects do. As the Iraqi landscape architect Jala Makhzoumi insists, "Fieldwork is a chance to understand the landscape. And landscapes are so complex, so layered. How else can you understand them?"[2] But fieldwork is not just an aid to understanding a landscape; it is a design tool. The Zürich-based landscape architect Günther Vogt instructs: "You must expose yourself, as a body, to this landscape. What are your impressions, your feelings? And then you can start to design with it."[3] In this book, I offer my own perspective on landscape fieldwork: the tools we can use in such an endeavor, the interpretations we can reach, and an appreciation for the design conclusions the practice can generate.

Using five concrete examples from projects I've worked on over the course of twenty-five years, this book shows landscape architects *how* to get out and fieldwork—and how to incorporate the concrete, experiential insights gained during that process into their own research, pedagogy, and design projects. Importantly, in encouraging landscape architects to go out and do fieldwork, the book makes a case for more collaboration and diversity in our approaches to understanding landscape. It brings disparate fields together, especially landscape architecture and anthropology, to activate the common but largely unsettled grounds between them. Many contemporary anthropologists, like landscape architects, are driven by concerns with climate justice and the protection of human dignity—and a shared commitment to a long and sustained embodied engagement with a site and its users. Anthropology is about understanding relationships, and what is landscape architecture if it's not about the construction of relationships?

Landscape fieldwork helps to bridge description and prescription, reflection, and action, and looks for meaning in design. Landscape fieldwork is a way of understanding the complexities of landscape through lived experiences and relationships with humans and nonhumans. Elements of a landscape such as flowers, fruits, plants, trees, soil, stone, gravel, environmental processes, and people—combined with time and space—together make up landscapes. Landscape architects reassemble these various components into novel physical forms. Landscape fieldwork integrates landscape architects' projective skills and tools for site analysis and design (such as drawing, measuring, photographing, remote sensing) with the ethnographic methods of anthropologists (participant observation, unstructured interviews, and writing reflexive fieldnotes). Through an ethnographic approach, we reveal landscape architecture's ethical and political agency to shape the world. When designers have a better understanding of peoples' values, dreams, and ambitions, surely, they can then be more successful designers. When the human aspects of landscape are acknowledged as much as the nonhuman, landscape can enter the realm of the political.

I have always preferred learning about what others have done, rather than what they instruct or what they think one should do. So, when faced with structuring this book on landscape fieldwork, I centered it on actual projects that introduce a range of approaches. Rather than write a didactic book with a chapter on how to conduct interviews, and another on how to take fieldnotes, I based *Landscape Fieldwork* on reflexive accounts of different forms of fieldwork, through five carefully chosen case studies differentiated by scale, topic, location, and design issues. By "case study," I refer to what Robert Yin calls the "teaching-practice case study," a format aimed at establishing a framework for discussion around a professional issue.[4] The aim is to offer a scaffolding for you, the reader, to learn from and debate, so that you will feel empowered to do your own landscape fieldwork.

The experiences described here inform one another to some extent, but landscape fieldwork is site-specific and needs to be adapted to every location. The sites range from smaller-scale public spaces to sacred groves, border landscapes, archipelagos, and wider regions. Exploring the landscape fieldworker's diverse skills, the book jumps between built and speculative landscapes, design processes, and landscape practices while traveling to sites around the postcolonial and Islamic world, including Western Europe, the Arabian Peninsula, the Caribbean, and Brazil. These cases offer a range of entanglements of people and landscapes that cannot easily be witnessed in Western contexts. This book goes all over the place in space and time, and I make no apology for that: landscape fieldwork takes complexity and connection as a starting point. Ultimately, this book demonstrates that experiential knowledge—gained from the embodied engagement of landscape fieldwork—can help to decenter Western canons of landscape architecture and offer new possibilities for the design imagination.

Reading the Field

For some reason, landscape fieldwork is largely absent from landscape architecture curriculums today. Landscape fieldwork is often embedded in teaching through field trips and site visits, but not discussed or formally conceptualized. Instead, the practice is seen as something landscape architects intuitively do—a semimystical and highly personal practice based in a set of ethics that cannot be intellectualized. Perhaps because it is done "outside"—outside buildings, outside academic conventions, and outside the predictable—is too easily dismissed as unimportant or unintellectual. Some colleagues have deliberately limited the use of landscape fieldwork, believing it a distraction from design creativity and the art of production. From a professional perspective, landscape fieldwork can also appear highly uneconomical in terms of the time and money it demands, which is more of a perception than a fact. Such attitudes ignore the profound insights that can arise from a deliberate and long-term engagement with a site and its inhabitants. I posit that this void stems most fundamentally from an avoidance of recognizing people as an inherent component of a landscape. Whatever the reason, this lack of instruction leaves landscape architects ill equipped for realizing the multidimensionality of the places and communities they are designing for.

Across the chapters of this book, I offer various ways of engaging with landscape fieldwork as a methodological practice. I apply the descriptive, participatory, and reflective aspects of ethnography to landscape architecture, and the imaginative, projective, and prescriptive capacities of landscape architecture to ethnography. I begin with the design and construction of a village square in my own hometown, revealing a landscape fieldwork process that is deeply grounded in the ecologies of that landscape. This early project introduced a question that I have grappled with for the rest of my career: How can I come to know the ecologies and inhabitants of an unfamiliar landscape as if it were my home? The second chapter explores one experimental method I practiced working for a firm in London: random sampling of a landscape as a way of trying to understand ecologies where humans and nonhumans coexist. This process offers insights but has limitations too. While the quantity of sites investigated can produce a "thickness" of description (to acknowledge a term popularized by Clifford Geertz), the process lacks a deeper long-term engagement with a site and its users that ethnography offers. The third case study introduces a more formal ethnographic approach to fieldwork. Through my year of fieldwork in Bahrain, we see the benefits of a classic ethnographic model of a fieldworker spending a year in the field, but with the landscape as the object of study, rather than a particular social group. We also see the big weakness of this process: ethnography works at a pace that is too slow for most design planning processes. The fourth chapter, set in The Bahamas, proposes that ethnography—and design interventions—can be done more quickly by working collaboratively. And the last chapter takes readers to Brazil to detail another approach to working within the limitations of time, through shorter bursts of fieldwork spread over a period of years, and fieldwork that at the time of writing is still ongoing. Taken together, the five chapters offer a central thesis: when landscape architects *understand* the landscape we are designing—including its users—we can more successfully work with a site's inherent complexity rather than try to simplify it.

The landscape historian/cultural geographer J. B. Jackson attests: "A rich and beautiful book is always open before us. We have but to learn to read it."[5] At the same time, the anthropologist Clifford Geertz tells us, "Doing ethnography is like trying to read (in the sense of 'construct a reading of') a manuscript—foreign, faded, full of ellipses, incoherencies, suspicious emendations, and tendentious commentaries, but written not in conventionalized graphs of sound but in transient examples of shaped behavior."[6] But how do we learn to read the complexities of a landscape? Landscape fieldwork helps us read landscape's open book.

Landscape fieldwork teaches us, importantly, how to *feel* a landscape. It's easy to dismiss emotional attachments and responses to a site as unobjective and unscientific—but landscape fieldwork embraces the subjective and relational aspects of a place as much as its objective function and appearance. Landscape fieldwork recognizes the forms of knowledge gained from the field—often collected through community archives, interviews, oral histories, sketches, and fieldnotes—as equally valid as scientific knowledge, formal reports, and official histories. Landscape fieldworkers must triangulate between the knowledge received from the field, the literature, and their own perspectives. It is almost impossible not to project our personal values on other humans and nonhumans, and it is essential to acknowledge the landscape fieldworker's centrality. Often, grappling with various forms of knowledge

Figure 4. Field in La Rochelle, France, 2020. (Photograph by Philippe Coignet, OLM, Paris)

can change our understandings and way of thinking—and disrupt the design and research process. That's because emotions often expose us to the unknown. But if the goal of landscape architecture is to serve users more fully, we can't do so without landscape fieldwork—that is, without an open-ended approach that has the space to be emotional *and* rational, qualitative *and* quantitative.

In that sense, landscape fieldwork is more than a research or design method: it is an ethos that has the potential to generate new knowledge and theories of site, unearth novel design challenges, and illuminate robust design solutions. Moreover, landscape fieldwork has the potential to add to theories of landscape architecture, in the process filling gaps in knowledge and expanding and enriching the discipline. Landscape fieldwork offers us the means to understand a landscape in and through material, spatial, sensual, social, and temporal experiences—that is, through the tactility and feeling of the field. This approach expands and revalues forms of experience and knowledge that have often been left out of the field. As we will see, landscape architects from different traditions use a range of methods that are rarely documented or conceptualized. Since most regions of the world have neither a formal field of landscape architecture nor written landscape histories or treatises, landscape fieldwork is essential for knowledge building if we are to break out of the Western landscape architectural canons and imagine alternative possibilities for the future.

Origins

The founding figures of the field of landscape architecture as we know it today and to whom we owe so much—such as Geoffrey Jellicoe, Gertrude Jekyll, Frederick Law Olmsted, and Roberto Burle Marx, and many others who are less well known—understood the agency of fieldwork. Each of these founders were first trained in another field unencumbered by a contemporary landscape architectural education. They each engaged in deep and meaningful fieldwork that gave them a powerful repository of knowledge that they drew upon throughout their landscape architectural careers. For the founders of the field, fieldwork was a framework with which to engage the aesthetical and ethical dimensions and the poetizing side of landscape. These founding figures engaged deeply with the land and, importantly, with the people who shaped the land. They methodically observed the world around them and used multiple forms of media such as notes, sketches, and photographs to document their observations. They recognized that the forms of knowledge generated from landscape fieldwork are essential for understanding landscapes and, in turn, for landscape architectural success. Despite the early influence of Jellicoe, Jekyll, Olmsted, and Burle Marx, however, the discipline has not fully built upon its landscape fieldwork roots, so it's important to briefly review their contributions here.

Jellicoe

Sir Geoffrey Jellicoe (1900–1996), the leading British landscape architect of the twentieth century, originally trained as an architect at the Architectural Association in London in the 1920s, where he studied the capital's buildings through field visits and sketching. But he credited his discovery of landscape architecture to his year of

Figure 5. Sir Geoffrey Jellicoe sketching in his garden, 1984. (Photograph copyright Tara Heinemann)

surveying Italian Renaissance gardens. Having won a *Prix de Rome* through the British School in Rome, Jellicoe spent a year in Italy together with his friend J. C. Shepherd. Jellicoe worked in the field measuring the gardens and writing descriptions, while Shepherd drew the plans and elevations of the gardens back in the hotel. For Jellicoe, the most influential practice he developed was the act of pacing—a form of land-surveying where measurements are coordinated to the length of one's step.[7] As Jellicoe stepped his way through Italian gardens, he became aware of the metrics of time and space that are so central to landscape architecture. Landscape is never static, and Jellicoe experienced the multidimensional and multisensorial aspects of landscape and a tactility that transcended other art forms. Consequently, Jellicoe decided to focus his career on the architecture of landscape, rather than the architecture of buildings, and became founding president of what is now the Landscape Institute (LI) in London and the founding president and later honorary life president of the International Federation of Landscape Architects (IFLA). For Jellicoe, landscape architecture had a much greater impact on the environment and a greater potential for the future of humanity than architecture, painting, or sculpture. Jellicoe concluded that landscape is the "mother of the arts."[8] Not only did Jellicoe's fieldwork lead to his conversion to landscape architecture, but it provided him with an intellectual and methodological basis for his future career. Many of his subsequent seventeen books and countless projects refer to this formative year.

Olmsted

Frederick Law Olmsted (1822–1903) worked as a surveyor, farmer, and journalist before eventually coming to be known as the "father of American landscape architecture." Olmsted was an exponent of landscape fieldwork. His early experiences

Figure 6. Frederick Law Olmsted Sr. seated, perhaps in the Hollow at his home in Brookline, Massachusetts. (Courtesy of the National Park Service, Frederick Law Olmsted National Historic Site.)

traveling in China and around Europe, including the British and Irish Isles, were influential and helped shape his fieldwork approach.[9] Before the US Civil War, Olmsted traveled in the American Deep South, documenting the horrors of slavery for *the New York Times.* Olmsted's firsthand accounts from the field were deeply ethnographic in the sense that they engaged with enslaved people and their forced interactions with the land. Olmsted's immersive and deeply engaging field-based written representations also foregrounded the environmental, political, and social aspects of landscape. Olmsted is perhaps best known for his design of immersive public landscapes. Central Park in New York City, which he designed with Calvert Vaux, continues to influence ideas of what landscape architecture is and might be the world over. The motivation for social justice and civic life was nurtured by Olmsted's fieldwork and manifested also in his other public landscape projects, such as Prospect Park in Brooklyn, and the Emerald Necklace in Boston.

Jekyll

Gertrude Jekyll (1843–1932) was a prolific plantsperson, garden designer, writer, and artist, responsible for designing over four hundred gardens and authoring two thousand horticultural articles and fifteen books. While it is tempting to think of Jekyll as the "doyenne of herbaceous borders,"[10] especially noted for her carefully chromatically curated drifts of plantings, she was, according to her biographer Sally Festing, much more than this. Jekyll was, according to Festing, "a major shaper of spaces."[11] Festing quotes Christoper Hussey's view that Jekyll was "perhaps the greatest artist in horticulture and garden-planting that England has produced."[12] Jekyll is also famous for her fruitful collaboration with the Arts and Crafts architect Sir Edwin Landseer Lutyens. Working together on numerous projects, Jekyll brought a visibility to garden design in the late nineteenth and early twentieth centuries. Perhaps Jekyll and Lutyens's most well-known collaboration is Munsted Wood, in Surrey, about thirty miles southwest of Central London. Munsted was Jekyll's home for thirty-five years. Lutyens designed the house, and Jekyll the garden. What is less well known about Jekyll, however, is that she trained as an artist—this perhaps explains the attention to chromatic detail in her work. In her formative years, Jekyll took long immersive trips abroad and recorded them with watercolors and sketches in her sketchbooks.[13] When she was just twenty years old, in 1863 Jekyll made her first

Figure 7. Gertrude Jekyll in her garden, Munstead Wood, photographed by her friend Mr. Cowley, editor of *Gardening Illustrated.* (Image courtesy of Future Content Hub.)

foreign trip to Turkey and Greece. She sketched and painted the exotic foliage and started collecting bulbs and seed pods. She wrote about this trip in her first book, *Wood and Garden* (1899).[14] She later traveled to the French Riviera and joined a sketching tour of Italy.[15] During the winter of 1873–74, Jekyll visited Algiers. Writing in his memoir, her nephew Francis Jekyll describes "Arab cemeteries, chalk-white against the vivid blue sky or mountains, courtyards splashed with crimson sprays of Bougainvillea."[16] Not only did these trips introduce Jekyll to foreign and exotic plant species, but the colors and experiences from the immersion in foreign fields were clearly a lasting influence on Jekyll and she drew from them throughout her long career.

Burle Marx

In Brazil, Roberto Burle Marx (1909–1994) started out studying painting at the School of Fine Arts in Rio de Janeiro. But after spending eighteen months in Europe, he began experimenting with using native Brazilian plants in the family garden, which was unusual because early in his career most Brazilian gardens utilized imported European flora. He went on to become one the most influential figures

Figure 8. Roberto Burle Marx on a field trip in Ecuador in 1974. (Photograph by Luiz Knud Correia de Araujo, reproduced courtesy of Luiz Antônio Araujo)

in landscape architecture, combining art and ecology in the construction of landscapes. Burle Marx believed that landscape architects have an ethical and aesthetic duty to contribute to the betterment of public environments. He took expeditions into the Brazilian interior to hunt for plants, and propagated many varieties at his *chácara,* or farm, adjacent to his home in Guaratiba south of Rio de Janeiro. Indeed, so great was his love for plants that he is reported to have enjoyed sleeping among them.[17] Burle Marx's expeditions exposed him firsthand to environmentally destructive practices that tragically persist to this day, and they also exposed him formally to the curves and sways of the landscape. He saw the profession as a form of activist practice through which he could affect wider societal change to protect native flora and fauna and help bring about more just and equitable cities. He sought to create beautiful spaces that also served a wider social purpose. Burle Marx designed a huge number of gardens and public landscapes over his lifetime, ranging from small plazas such as the Praça Terreiro de Jesus in Salvador da Bahia, to the Largo da Carioca, Flamengo Park, and the iconic Copacabana Beachfront in Rio. All these projects were fundamentally informed by landscape fieldwork.

Landscape fieldwork builds upon a long-established dialog between landscape architecture and environmental planning, and ethnography and cultural geography that was especially strong at the University of Pennsylvania under the leadership of Ian L. McHarg and later Anne Whiston Spirn. Anthropologists including Dan Rose, Setha Low, Yuhedi A. Cohen, and others, shaped a generation of landscape architects in relational observation and design. This pedagogy straddling design, planning and cultural geography—so central to landscape architecture and yet liminal in terms of pedagogy—was also manifest at Harvard University with the likes of J. B. Jackson, John R. Stilgoe, Richard T. T. Forman, and others. This "thick" form of research also was evident in research initiatives such as the Urban Field Service that existed in the late 1960s and early 1970s.

Landscape as Feel Work

To get a better sense of the value of fieldwork in a landscape architectural process, let us briefly visit a garden in the south of England at Shute House. Shute is a refined

Figure 9. Ian McHarg (*center left*) with students on a studio fieldtrip ca. 1983 with Robert Giegengack (*center*), professor of earth and environmental studies. (Photograph by Frederick R. Steiner)

landscape, crafted from the Wiltshire countryside and fed by an ancient spring. The garden is a masterpiece of Jellicoe, who worked on Shute over many years from 1969 to his death in 1996. Jellicoe spent many weekends on the site with the original owners, Michael and Lady Anne Tree, exploring their minds.[18] Lady Anne had come from Chatsworth House where Lancelot "Capability" Brown had shaped the grounds, and she had, as Jellicoe put it, "the sense of landscape right in her bones."[19] I first visited Shute in October of 1995, after getting to know Jellicoe in the mid-1990s. The house and garden had by then been acquired by new owners, Sir John and Lady Suzy Lewis, who welcomed me warmly.

After spending a couple of hours exploring the garden on my own, walking and sketching and feeling increasingly exhilarated, Lady Suzy sensed that my emotions were deeply stirred. She generously invited me to spend the night in the guesthouse. Lady Suzy, I learned, was a daughter of the landscape gardener Esther Merton, and hailed from the Old Rectory at Burghfield House, where Merton had transformed the gardens, so she has the sense of landscape in her bones too. The guest quarters had direct access to the garden, and after dinner I went for a walk. The garden is made up of eight parts, each with a different character, size, and form, described by Jellicoe as "seven or more green compartments having been devised for the romantic, semi-watery wilderness that lay behind the wall."[20] They are the Canal, Temple, Temple Garden, Bog Garden, Stepped Cascade, Apple Porch, Rill, and the Grottos. Some are enclosed with high laurel hedges. Some are formal with waist-high box hedges.

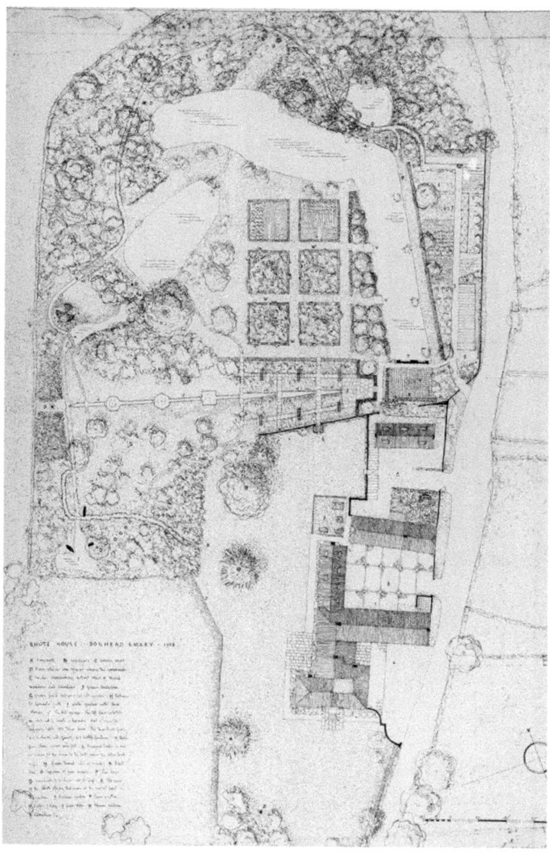

Figure 10. Shute House, plan, 1978. (Museum of English Rural Life, University of Reading, Geoffrey Jellicoe)

Figure 11. Shute, terrace. (Photograph © Sabina Rüber)

Some are informal, looking over the Wiltshire landscape. The disparate parts are held together with water in various forms, leading from the pool that is the source of all the water in the garden, feeding the canal, streams, pools, and the rill. Indeed, the spring is also the source of the River Nadder, which winds its way through Wiltshire for a further fifty-five or so kilometers (roughly thirty-four miles). Jellicoe saw it as a mix of classical and romantic styles, which together added up to what he termed a "cosmic" landscape.

That night was dark and moonless, and the air was thick with the fragrance of autumn. As I slowly made my way around the garden, I came to rely on my senses other

Figure 12. Shute, the box garden. (Photograph © Sabina Rüber)

than sight. Each area was a different experience. Parts of the garden were so enclosed that it was pitch black, and all I could do was listen, feel, and smell my way through. As I walked from compartment to compartment, I came upon the harmonic cascade, a series of water fountains in a rill, designed to replicate the sound of tenor, alto, soprano, and bass. The sound of water in the dark helped attune my body to the garden.

Visiting the garden that autumn night remains one of the highlights of my life. To this day, when I encounter the pungent smell of decaying apples, it reminds me of this nocturnal ramble. In my fieldnotes, completed on the train back to London the following day, I wrote: "I walked the garden in the dark. Statuary and forms of leaves and areas of total darkness elicited momentary fear. The sound of water cleansed the system."[21] It's true that on more than one occasion, I was anxious in this dark space that was unfamiliar to me. Maybe my nervousness made the experience even more memorable. I had to feel my way around and all the while try not to fall into the water. Although at the time I didn't realize Jellicoe intended the garden to be experienced at nighttime, I later found a passage where he describes the Temple Garden as "a place where a child might be slightly frightened. . . . Not a place for the timid on a dark night."[22] Clearly, Shute at night brought out the timid child in me.

Figure 13. Shute, the rill. (Photograph © Sabina Rüber)

Jellicoe intentionally sought to appeal to a visitor's subconscious. He wanted people to experience landscape at a deeper level than what they can see on the surface. To get there, beyond the visible, Jellicoe spent long hours immersed in the site—sketching, doodling, measuring, writing notes, and talking with the contractors and the gardeners. He constantly refined the design over the years; the innovative concepts were built after much thought and trial and error.

Jellicoe also incorporated physical references to several historic gardens and features, most obviously Rousham Cascade by William Kent, the eighteenth-century

Figure 14. Shute, the rill. (Photograph © Marianne Majerus)

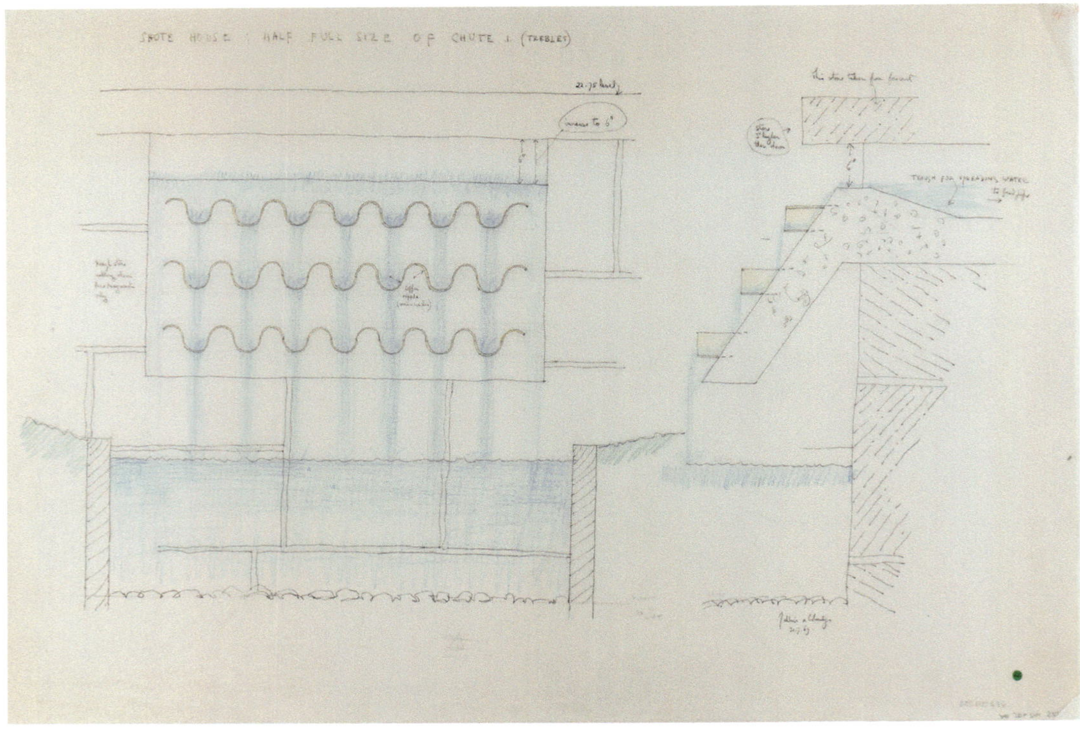

Figure 15. Water Chute 1 (*left*) and Water Chute 2, 3, 4 (*right*). (Museum of English Rural Life, University of Reading, Geoffrey Jellicoe)

landscape gardener. These were inspired by Jellicoe's fieldwork. Because of the central role of water at Shute, the historian Michael Spens, who has written several books about Jellicoe, most notably *Jellicoe at Shute,* draws parallels with the Villa Lante at Bagnaio, a garden from the Italian Renaissance gushing with fountains and rills and water jets, as well as the Villa d'Este at Tivoli.[23] However, when I did a study tour of Italian Renaissance gardens following in Jellicoe's footsteps, I saw greater parallels between Shute and the Villa Gamberaia at Settignano, with its terrace and compartmentalized sequence of spaces. Jellicoe considered the Villa Gamberaia one of the finest gardens in the world. There, he found a place for the different human

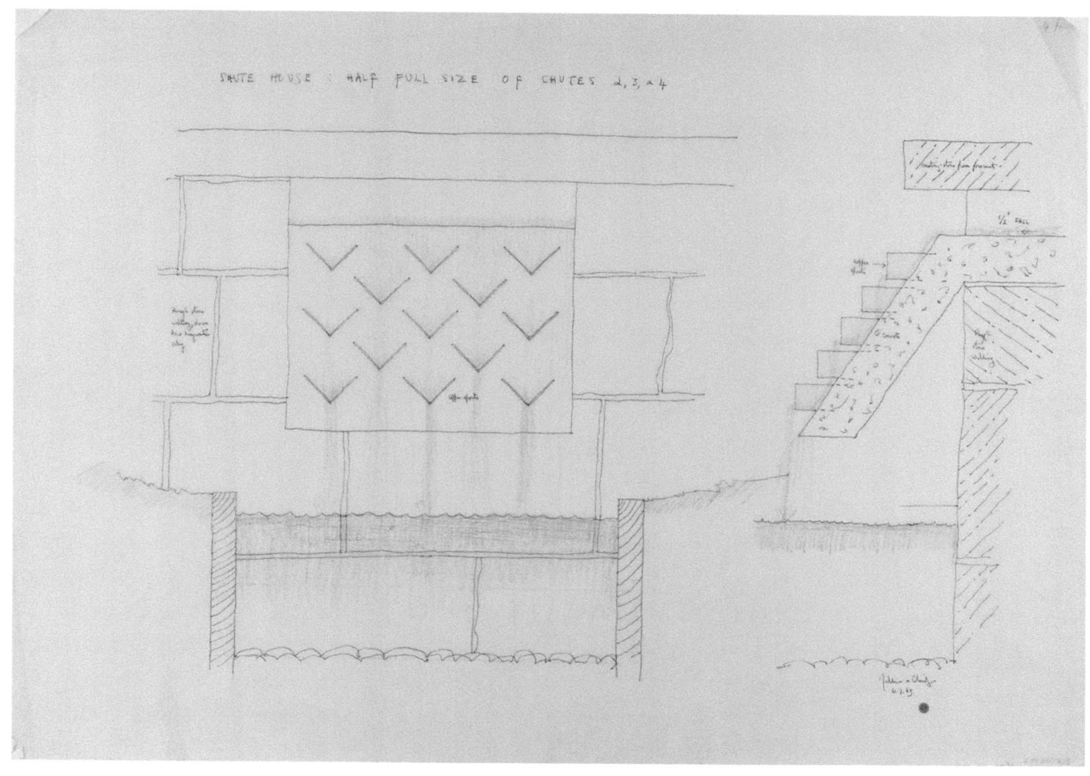

emotions. In 1927, forty years before he started working on Shute, Jellicoe wrote of the Villa Gamberaia: "There is a place for every mood." Jellicoe identifies melancholic spaces where Shakespeare's Hamlet would feel at home, and playful spaces for a prankster, and a common space for them to meet. "Hamlet will find an answering chord in the twilight of the bosco, mysterious, elusive, fantastic with the shapes of ilex; the joker can go and joke among the water steps and grotto; and the two can agree to differ in the most delightful of lemon gardens."[24] At Shute, the lemon gardens are adapted as an apple orchard, and the joker can frolic along the harmonic cascades. Clearly, Jellicoe's landscape fieldwork from the 1920s was still influencing him seventy years later. Indeed, his fieldwork also shaped his monumental and unfortunately named *Landscape of Man: Shaping the Environment from Prehistory to the Present Day,* which he coauthored with his wife, Lady Susan, who accompanied him on his field visits and photographed the projects.[25]

M. Elen Deming and Simon Swaffield, when speaking of ethnography and related forms of qualitative social research write that they "require the researcher to engage with the research as a feeling person, rather than as a detached observer."[26] Through feeling, landscape fieldwork brings us to understand the landscape behind, beside and beneath that which we see. To create landscapes to be felt, we first need to be able to feel. Fieldwork is also "feel" work.

Jellicoe summarized his creative design process in a diagram that outlines the relationship between client (A), client's instruction (A1), designer (B), designer's initial concept (B1), their joint subconscious (Y), the deep subconscious of the designer and the client (Z), as well as the final design project (X).[27] For Jellicoe, the client's instruction and the designer's initial concept converge at their joint subconscious and result in both the final project and the deep subconscious. As with any diagram, the processes are overly simplified. But it is significant that the deep subconscious is somewhere other than the project. Although Jellicoe does not say so explicitly, he achieved this joint meeting of subconscious minds through a deep and intimate engagement with the site and with the client. Jellicoe writes, "Like the portrait painter, the landscape designer needs to be a psychologist first and a technician afterwards. He needs to dig into the subconscious."[28]

At Shute, Jellicoe's long-term engagement and hands-on field-based strategy worked extremely well. The emotional resonance of the garden at Shute could not have been achieved without Jellicoe's multiple visits and meaningful interactions with the garden's users over an extended period. I can easily imagine Sir Geoffrey during these pivotal weekends at Shute with Lady Susan. They may have taken the train from London's Waterloo station on a Friday afternoon, been picked up at Tisbury station nearby, and arrived in time for drinks on the terrace followed by supper with the clients, the Trees. Sharing meals together is a time-tested way of getting to know, and understand, one's interlocutors. The conversation, probably centering on their shared passion for art, would have been augmented with the symphonic sound of rushing water. Sir Geoffrey said the harmonic cascade and rill was his favorite

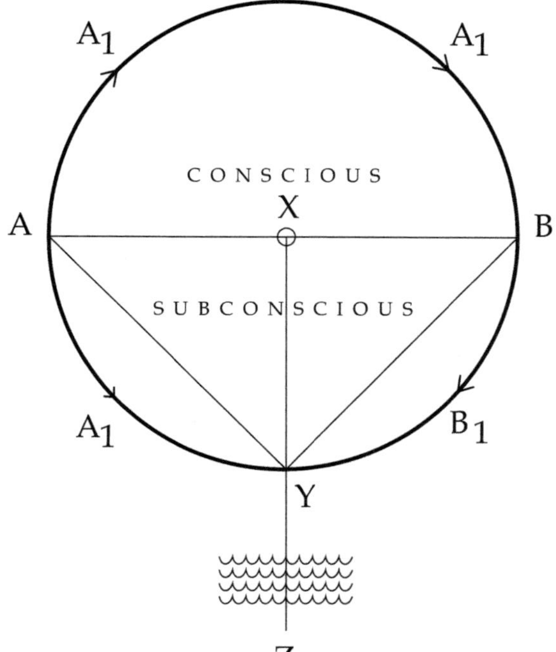

Figure 16. Jellicoe's diagram of the creative process. (Redrawn by Junho Kang after *Jellicoe at Shute*, p. 91)

work ever.[29] He so loved the site that he chose the garden at Shute as one of his last experiences of this world. After having supper with the Trees, he was driven back to his London apartment and vowed to never willingly leave it again. He remained there till his declining health forced him to retire to a nursing home.

Across the Fields

While the garden at Shute is a classic example of landscape fieldwork leading to a built design, increasingly landscape architects are being called to work in other kinds of settings. Landscape fieldwork is not just done out of doors, as John R. Stilgoe[30] and Tim Ingold suggest.[31] Landscape fieldwork is performed inside too. Landscape fieldwork can be done in coffeeshops, in dining rooms, or on terraces over dinner.

It may take researchers into archives and libraries, as we investigate and interrogate our sources. Landscape fieldwork may involve browsing satellite images or online platforms, not just as a complement to being in the field, but as part of it. Often, it involves conversations with people we may know, and with those we have yet to meet. Landscape fieldwork also continues long after site visits and conversations end, as the fieldworker interprets what they experience. Across these various spaces and forms of practice, the landscape fieldworker is driven by an obsessive attitude to leave no stone unturned. This endless fascination with doing whatever it is that you need to do leads to understanding a landscape more fully.[32]

That said, fieldwork means different things to different fields. Disciplines have developed their own fieldwork methods to understand, interpret, and intervene in a landscape. Here, I trace some of the methodological similarities and differences across a selection of fields, looking at the data collection methods, goals, tools, and scales. In each case, I show how fieldwork shaped that field. I take the cognate fields in alphabetical order beginning with what the anthropologist Tim Ingold terms the "four As"—anthropology, archaeology, architecture, art[33]—and then move on to ecology, gardening, geography, history, and urban planning. This is not an exhaustive list by any means, but merely an illustrative sample. Landscape architecture, as a multifarious discipline, borrows elements from all these fields, and more.

Anthropology

Anthropology, perhaps more than any other academic discipline, depends on knowledge gathered in the field. The Department of Anthropology at Harvard University describes anthropology as "the study of human diversity in the distant past and the present and teaches us to recognize the remarkable array of circumstances in which human beings live their lives and make meaning from them."[34] In order to understand these meanings, anthropologists will typically spend at least a year in the field, doing fieldwork: living among a community, building trust, learning language and codes and patterns of behavior, and carefully and methodically noting details not only of peoples' daily lives but also aspects of their relationships to objects and the environment, and to one another. "I did not know much about anthropology, but I lived it," declared Laurie Olin, the eminent landscape architect, of his upbringing in Alaska, highlighting the fact that anthropology is about active, lived experience.[35]

Anthropologists can use many tools, such as cameras, voice recorders, microphones, and video recorders. Still, their core tools are paper and pen, and (increasingly) smartphone and computers, for writing and interpreting their fieldnotes. Anthropologists keeps detailed notes—scratch notes, fieldnotes, and "headnotes"—which they review again and again as they craft their eventual ethnography.[36] There are many types of anthropology, and here I refer to sociocultural anthropology.

If anthropologists were parachuted in to do fieldwork at Shute, they would obviously talk with the owners and their family, but their training which often favors the subaltern, might incline them to also spend time with the gardeners and, if they could find them, the laborers that planted and built the garden. They may live for an extended period in the adjacent village of Donhead St. Mary, or one of the nearby settlements of Donhead St. Andrew, Ludwell, or Shaftesbury, and participate in everyday life with the staff and neighbors. They may even ask for a job at Shute and work on the property in their process of participant observation. Depending on their guiding research questions, anthropologists might also analyze various media and social media to see how people and spaces are represented. They may be more interested in the processes that led to the building of the garden and its maintenance than in its eventual built form. Or, they may be fascinated with the nature and culture relationships, as much as the garden being a representation of privilege within a hierarchical social structure.[37] Anthropologists might look at census data, perhaps through the Wiltshire County Archives,[38] and find the current population of about four hundred is less than half the nine hundred recorded in 1841. Anthropologists interested in changing social relations might be interested in the many reasons for this drop, including the way statistics are gathered. They might also be interested in larger flows through Shute, such as the origins of materials and the roles of the property within global flows of capital. Whatever their initial research questions, a defining feature of anthropological fieldwork is that the eventual project would strive to be shaped by what emerges from the field, rather than preconceived notions they bring to the field.[39]

Archaeology

Archaeology, like anthropology, is heavily fieldwork-dependent, perhaps even more so, because archaeologists often literally dig into field sites. Indeed, it is sometimes considered a subset of anthropology. Through digging, archaeologists find objects, artifacts, and physical remains which they analyze to understand patterns of human history and more usually prehistory.[40] Archaeologists will look for the concealed structures in an ancient field or garden that may infer how people experienced and interacted with one another and the world around them. Like forensic scientists, archaeologists understand situatedness, a form of thick description, or thick understanding, that acknowledges and describes experiential relationships between people and the spaces around them.[41]

Because archaeologists literally dig into the field, the trowel is their most basic tool. The archaeologist's trowel is supported by shovels, brushes, and sieves, as they slowly and methodically analyze the earth looking for remnants. Some archaeologists will work with tiny blades and picks for microexcavations as they search for human or animal remains. They may use a theodolite for site surveying and geographic information systems (GIS), satellite imagery, and remote sensing for working at the larger scale. Like anthropologists, archaeologists will record their observations through detailed and methodical fieldnotes, writing, sketching, drawing, and photographing. Archaeologists differ in that their work usually tends to be more spatially defined than that of anthropologists, requiring precise spatial measurements. As a result, the archaeologist consults and draws maps and plans. Having said that, there is also a temporal distinction—that is, archaeologists study the past and anthropologists the present.

An archaeologist at Shute would likely want to know what was on the site a very long time ago, probably before the garden existed, although historical archaeologists are more interested in recent history. Archaeologists would consult historical records and combine these with a detailed site survey using maps, compass, and a GPS device. Subject to the proper permissions, they might dig, if they thought there was sufficient evidence to merit a dig. While archaeologists would know that Shute is a former rectory, they might be more interested in the so-called Roman spring which Jellicoe diverted to form the series of water features in the garden at Shute. They might be equally interested in the Roman Road from Bath to Badbury Rings that crosses the parish passing close to the nearby St. Mary's Church; the road would have facilitated agriculture and commerce, and artefacts revealed from excavations adjacent to the road suggest the area was prosperous during the Anglo-Saxon era

(410–1066). The archaeologist may be interested also in the ancient earthworks at Winklebury Hill which contain an Anglo-Saxon cemetery. Going deeper into ancient history, the Mesolithic microliths, a Neolithic long barrow, bowl barrows, and ditches and an Iron Age hill fort near the edge of the civil parish might also pique the archaeologist's curiosity. And of course, the archaeologist will know that Shute is about twenty miles (thirty-two kilometers) from Stonehenge. The archaeologist might not be all that interested in the physicality of the contemporary garden, but instead be curious about what lies beneath, beside and beyond the garden, and what existed there a long time ago.[42]

Architecture

Architecture and landscape architecture share much in common in terms of their spatial and projective capacities, always imagining alternative futures. Although both fields are concerned with how we will live in the future (designer) as much as how we live now (anthropologist) and lived in the past (archaeologist), architecture is usually concerned with an object situated in space, rather than a living horizontal system.

Since architects are concerned with space and form, they will instinctively draw, sketch, paint, doodle, photograph, and maybe take notes. Their tools are often rudimentary, a notebook and pencil or pen, or some other sketching device. Their notes and drawings are spatial and projective. An exemplary albeit unusual case of architectural drawing is the work of Kaijima Momoyo, cofounder of the Tokyo-based architecture firm Atelier Bow-Wow. Kaijima, who is also a professor of "behaviourology" at the ETH in Zürich, has devised a remarkable graphic style that depicts buildings as they are used and might be used in the future. While architectural conventions reveal the raw form of individual structures out of context, Atelier Bow-Wow imagines spaces in everyday life. As well as sketching, drawing, painting, and writing, architects engage with photography, film, sound, and other forms of media. And those engaged with an activist practice may work with chisels, saws, and hammers to construct buildings.

Depending on the design attitude of the architect, an architect at Shute may first want to know what sort of new building or addition the client wants. The architect will know that the L-shaped Shute House originates in the late sixteenth century, with an early seventeenth century addition to the south. The front door is said to come from Bowood House, now a hotel, spa, and golf course, whose grounds were originally landscaped by Capability Brown. Architects may consult books such as Pevsner's *Buildings of England,* or agencies like the Wiltshire Historic Buildings Trust.[43] Perhaps the architect's work will be more about renovation, or conservation, or in making the ancient building more energy-efficient, but architecture will by and large focus on structures. The architect may propose a cabin in the garden, or a gazebo, or an extension to the house, but whatever is offered, convention suggests it will likely be bounded by walls and a roof.

Art

The artist may take many approaches and use many mediums, whether they be visual, sensual, phenomenal, conceptual, or written, or all of these. Few writers, for example, are not inspired by landscapes of one form or another, and they translate that landscape into words.[44] Let's focus on the work of visual artists and a specific type of landscape painter, the plein air painter.

The basic tools of the plein air painter typically include a collapsible easel, a paintbox and tripod, folding stool, clamp-on umbrella for shelter and shade, paints, brushes, palette knife, solvent and painting medium, wet panel carrier, paper towel or rags, small plastic trash bags, and bungee cords, according to Ross Merrill, artist and a former chief curator of conservation at the National Gallery of Art in Washington, DC.[45] David Hockney cites a Chinese saying that there are three tools for a good painting: "the hand, the eye and the heart." Two out of three won't do; all three must be present.[46] Hockney is talking about the technical skills of painters, their perception, and their passion.

The artist at Shute might set up an easel like Hockney and work quickly and methodically in between showers of rain or changing light conditions. The work may capture the grandeur of the landscape, including the land and sky, and depict more of the field than is immediately shown in the image. The artist may work quickly and complete the painting in situ, or complete the project back in the studio. Nevertheless, the painting itself is the product of a protracted period in the field, incorporating prior sketches and field observations, not to mention the past experiences of the artist. The artists may, like Hockney, keep sketchbooks, which are then used back in the studio to complete the painting. It's important to recognize that the artists' fieldwork is often initiated before the site visit and extends beyond a site visit. Whatever, the subject of the painting, the aim is to take the beholder beyond what the eye can see.

Ecology

Ecology is the dynamic study of change, and how a landscape may develop over time driven by human land-use conditions and climate modifications. The tools of the ecologist are plentiful and increasingly sophisticated and depend on the specialization of the ecologist. Tools might include an altimeter, a clinometer for measuring angles and tree heights, a drone, environmental DNA for genetic testing (eDNA), field guides, field glasses or binoculars, handbooks, image analysis (where different trees reflect different spectrum which are then measured), a light meter for measuring light intensity, a meter stick, Munsell Soil Color Chart Basic Set (which is brought to the lab for chemical analysis), a pH test kit, and a planimeter for measuring distances on maps.[47]

The ecologist at Shute might well be interested in the biodiversity of the garden and its impact on the surrounding landscape and urban settlements, and vice versa. The ecologist may focus on one aspect: the peculiar ecosystem of the garden, and how it may be improved. The ecologist might be interested in the ecological corridors—the hedgerows that are a distinctive feature of the English countryside—that link Shute with the surrounding Wiltshire landscape. Ecologists may focus on biodiversity conservation and understanding the local vegetation, wildlife, and their manifold relationships. Urban restoration ecologists, such as Steven Handel, might be interested in how to restore the ecosystem to an improved or optimal state and will use historic records of biodiversity, dispersal potential (i.e., what can come into the site), and tests of physiological tolerances to the local conditions.[48] In all cases, the ecologist will be concerned with the relationship of the local landscape to the wider biosphere, and vice versa.

Gardening

Gardeners are the ultimate fieldworkers, spending most of their time in the field—hail, rain, or shine. They directly apply their field-based knowledge in the field. This is often tacit knowledge; a triangulation between what the books say, what will grow in a specific location, and design intention.

Gardeners use many tools. Like archaeologists, they will likely have a trowel or spade. They will also have secateurs or pruning clippers, a garden fork, perhaps a pH soil test kit, a spade, a shovel, a watering can and sprayer, a rake, and even a wheelbarrow. They will have gardening gloves, of course, and gardening clothes, including shoes, hat, and raingear or sunblock. Gardeners will have a range of books to guide their species selection, and almanacs to keep track of weather predictions, and the lunar cycle. They will usually have pencil and notebook or some drawing apparatus, such as an iPad, as they note growing habits, colors, flowering period, and adjacencies. They often keep a diary and document daily happenings, such as "First baby rabbit of the season," or "First daffodil," and record the idiosyncrasies of their location and the microclimates within the garden. Gertrude Jekyll, the "doyenne of herbaceous borders," kept travel notebooks. Like Burle Marx, Jekyll was trained as an artist, driving her great interest in color.[49]

Gardeners at Shute will be interested in the planting plans, and what they can infer with the juxtaposition of different plants with one another and in different light and weather conditions. They may be interested in the politics of native plants versus imported species. They may be curious about the unusual play with plans, the waist-high box hedges, and the Brazilian giant rhubarb (*Gunnera manicata*) with its huge leaves that fill with rainwater and tilt over and empty into the rill. At Shute, Lady Anne Tree selected the plants, and the planting plans were carefully redrawn for Jellicoe. Usually, Susan Jellicoe would choose the plants for her husband's gardens. Jellicoe himself admitted to knowing "nothing" about plants and trees although he insisted that he knew about the "spirit" of these things, which he felt was more important.

Geography

Geography is the writing about the land, as ethnography writes about people. But this distinction is not terribly accurate, for geography includes people too. The human geographer Yi Fu Tuan describes geography as "the study of earth as the home of human beings."[50] The *National Geographic* defines geography as "the study of places and the relationships between people and their environments."[51] Whatever geographers study, they will likely try to understand a global phenomenon or topic as it acts and is understood through a specific place. In this case, geography—like landscape architecture—is not tied to any one scale but is concerned with relationships among scales. The tools of the geographer includes maps, globes, charts, data, GIS, and other more specialized equipment to understand humans' relationships with the earth and vice versa.

The geographer at Shute might take any number of approaches, but it will be rooted in the place and its relationship to the world. A geographer interested in climate change may, for instance, be curious about the garden's sustainability indicators and the impact of changing climates on the surrounding Wiltshire landscape. The geographer may want to know the degree to which the use of resources is linear or cyclical: cyclical use implies recycling and the use of renewable energy and resources. Geographers will likely be interested in human interactions, in the source of materials, in the role of the local community as much as in the role of the garden in promoting ecological biodiversity.[52] They might also be interested in globalization as demonstrated through the selection of garden plants, or the fractured geographies and temporalities of the various garden styles represented in the garden from the English Romantic to the Italian Renaissance.

History

"One of the great things about history, is that everything has a history," said Ann Blair, chair of the History Department at Harvard University.[53] Since historians study everything, and we're focused here on a garden, let's focus on garden histories. John Dixon Hunt, the garden historian, tells us that history is "the narration of past events."[54] He goes on to tell us that history needs a narrator and an audience and that "a narrative of the past implies or offers something for its present-day listeners, which, further, implies a future—by way of some guidance, instruction or emulation of the past for the years yet to be."[55]

Historians will use documents, drawings, objects, images, photographs, to unravel the complex narratives of a site and its context. Often limited in archives to the use of a pencil and paper, the historian's tools are mostly intellectual. Along with the pencil and paper, historians use ideas. Historians will look for chronology, causation, evidence, perspective, and skepticism.[56] They may also measure, use photography, and use 3-D scanners.

Jellicoe, who in 1925 coauthored *Italian Gardens of the Renaissance,*[57] based on a year of fieldwork,[58] was profoundly aware of landscape architecture history. He cautioned, "I look upon history solely from the point of view of what can it point to the future. And I feel this is extremely vital for people to know."[59] Historians at Shute will want to focus on some specific aspect of Shute's history and its relevance for the present and the future. They might begin their fieldwork in archives, such as the Wiltshire and Swindon History Centre in Chippenham[60] or the Dorset County Archives.[61] They might also visit the Geoffrey Jellicoe Collection at the University of Reading's Museum of English Rural Life,[62] or the Garden Museum in London.[63] The Susan Jellicoe photographic collection at MERL offers visual documentation for much of Sir Geoffrey's work and may offer clues for how the gardens of Shute matured over the years.[64] Alternatively, historians might be interested in the garden's role in the development of Jellicoe's work, or its position in English garden history, or the garden's position *vis-à-vis* garden history more generally. They may compare Shute with gardens elsewhere, such as nearby Stourhead, or the Villa Lante and Villa Gamberaia. The historian may even be interested in the wider Wiltshire landscape, that includes Shute. But whatever the topic of study, the garden or landscape historian will be interested in the implications for today, and tomorrow.

Planning

"Plans require context and vision to provide the connection between what we know and what we want to do" urges Frederick R. "Fritz" Steiner.[65] There are many types of planners: city, ecological, environmental, regional, and urban planners. Given the plurality of planners, they use multiple site-specific tools and methods. Planners may engage in a range of analytic methods, such as William H. (Holly) Whyte's time-lapse photography of small urban spaces in New York that resulted in his acclaimed film and book, *The Social Life of Small Urban Spaces.*[66] Whyte's analyses of patterns of human behavior from photography and films influenced generations of designers and planners, including the London-based office Space Syntax. Yasser Elsheshtawy uses similar observational methods to analyze the urban spaces of low-income migrant workers in Dubai and Abu Dhabi; his unnerving images show the city's backstage, how cities are inhabited, over and above more familiar formal representations. We see the human side behind what Rahul Mehrotra terms the "landscape of impatient capital."[67] Whatever the approach, planners work with communities to understand their needs and concerns, working with statistics and quantitative data to understand patterns of human interaction, draw up plans, and write policy documents.[68] In addition to documenting the results of time-lapse photography, planners may turn to spreadsheets, GIS, building information modeling (BIM), SketchUp, and Google Maps.

The planner at Shute may be interested in land use, parking spaces, real estate values, and the road widths, as well as the local codes. How many visitors could this garden accommodate, especially given the very narrow roads that lead to the site? A planner may consult the county plan or a development plan of the area, perhaps even draw a section though it, like Patrick Geddes and the Valley Section, where Geddes traced the path of a river and the various occupations and tools associated with the section. The planner may consult the Wiltshire Council's Planning Policy,[69] and the Wiltshire Planning and Building Control Public Register.[70] The planner may be involved with a statutory public consultation process. A planner going into the garden at Shute might also be interested in its relationship to Donhead St. Mary and the nearby villages of Donhead St. Andrew and Ludlow. A planner may understand Shute's role in providing employment and in the physical infrastructures required to support this garden. Simply put, a planner would be concerned with how Shute fits in as a piece of a greater environment or larger landscape to be managed and planned.

What Is Distinctive to Landscape Fieldwork

Landscape architecture originates in multiple genealogies and, as a result, draws from all the above fields and more. We might easily include art history, botany, engineering, and sociology in the mix. While there is much we can learn from other fields to apply to landscape architecture,[71] landscape architecture is already a multidisciplinary hybrid consisting of multiple assemblages of relationships. The fields themselves are hybrids and are further hybridizing into fields such as situated urban political ecology, for instance. Each of the fieldwork-based fields listed above offers different tools and ways of understanding a landscape—but in practice, within landscape, the disciplinary lines often break down.

One subsequent visit that I made to Shute for landscape fieldwork illustrates this point. During a lengthy and enjoyable unstructured interview with the owners, I learned many new details about the landscape. The owners showed me the ancient hole for almsgiving in the wall of the house, a former rectory. I learned of the myriad challenges of garden maintenance and in keeping a straight line on top of a high yew hedge, and the woes of box blight. I learned of the reported healing power of the gardens, where the spellbinding sound of water washes ailments away. Moreover, I learned that the "Roman" spring that fills the garden with water and the source of the River Nadder is believed to have fertility properties. The spring's qualities are well known locally and have been for generations. To this day, people attest to the spring's powers and derive benefit from them. Whether the "Roman" in "Roman spring" is a reference to gods or not, I was surprised at how this important information is omitted from the literature. Nowhere in the copious writings on Shute have I found this detail mentioned. As the source of all the water in this garden, the magical power of the spring's water seems a crucial factor. Because of his lengthy engagement with the garden, Jellicoe must have known about this quality of the spring but chose not to write about it. Despite his early foray into fieldwork, perhaps Jellicoe was hesitant to include unorthodox details such as this, fearing critique among the intelligentsia. Or, more likely perhaps, this was part of Jellicoe's subconscious brief, a deep secret behind that lies behind the garden's magical power. Landscape fieldwork confronts us with such unknowns, that often confound disciplinary boundaries, and open us to their possibilities.

As landscape architects reflect on the field, we need to continue looking outward to expand into new methods and perspectives, and looking inward toward what defines us. While landscape architects must continue to develop our own methods and tools of analysis, as with any specialization and desire to define exactly what landscape architecture is and does, there exists the potential to reduce and restrict, as opposed to opening. Elizabeth Meyer writes about the possibilities for opening new possibilities and interdisciplinary spaces when she writes, "New directions for site practice might look less at new tools for how to read sites and more at finding spaces within which to imagine sites. For those spaces might be as much between disciplines as they are between surfaces, membranes, and operations."[72] With this in mind, let us look at the essential components of landscape fieldwork, while at the same time keeping them open enough to allow for more possibilities.

How we come to understand landscapes is splendidly documented elsewhere by Carol J. Burns and Andrea Kahn, James LaGro, Kevin Lynch and Gary Hack, Elizabeth Meyer, and many others. Landscape fieldworkers are familiar with site analysis and sitework—typically understood as processes for preparing a site for a particular project or activity, such as excavation work or the grading of the topography. What,

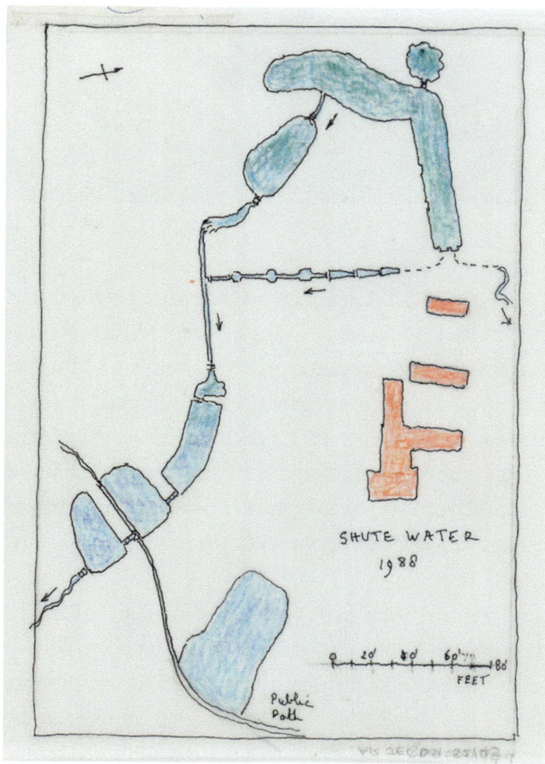

Figure 17. Shute, waterscape. The spring is on the top right. Sketch by Geoffrey Jellicoe, 1988. (Museum of English Rural Life, University of Reading, Geoffrey Jellicoe)

Figure 18. Shute, ancient spring. (Photograph © Marianne Majerus)

then, is the difference between "site" and "field"? If site is the location of something, the field is the ground on which the site is located. It is important to keep this distinction in mind when it comes to the sitework versus fieldwork. Andrea Kahn and Carol J. Burns point out that landscape architecture regards site as "material terrain," while other fields have different understandings of site. At the same time, they highlight the fallacy in thinking that a site is a straightforward entity and suggest that the definition of site is dependent on multiple factors, not least those who hold power over it.[73] The fieldworker, by contrast, will work out protocols after the "site analysis" rather than before.

The landscape fieldworker goes into the open field with questions, an informed intuition or inkling, and then allows the fieldwork to inform the result. The British anthropologist Marilyn Strathern writes about anthropological fieldwork in these terms: "What research strategy could possibly collect information on unpredictable outcomes? Social anthropology has one trick up its sleeve: the deliberate attempt to generate more data than the investigator is aware of at the time of collection. Anthropologists deploy open-ended, non-linear methods of data collection which they call ethnography; I refer particularly to the nature of ethnography entailed in anthropology's version of fieldwork. Rather than devising research protocols that will purify the data in advance of analysis, the anthropologist embarks on a participatory exercise which yields materials for which analytical protocols are often devised after the fact."[74] Strathern reminds us that what might seem tangential at one stage may reveal itself to be a central key to understanding some phenomenon afterward.[75] As landscape architects engage more with urban issues, it is imperative that we cultivate this kind of openness as well, including a receptivity to engaging more with people.

Getting Started

When faced with a new project, the landscape architect Günther Vogt will look in the cupboard to decide what tools will be needed for a particular landscape. "Are we doing photos? Are we doing sketches? Sometimes we have several tools we are working with.... It's about measuring the landscape," Vogt reports.[76] The tools of the landscape fieldworker are too numerous to list here but include all those listed above and will involve tools of measurement, immersion, recording, and imagining. Above all, the most central tool is the human body. It is through embodied, immersive experiences that we come to understand a landscape.

First, one might rightly want to ask: What *is* landscape? But that question will take us down a slippery slope, for there are many definitions of landscape.[77] The landscape that is the object of landscape architecture, for instance, may be considered as a refined form, in the same way that literature is a refined text.[78] For some, landscape architecture is a Western invention, a patriarchal spatial practice for the elites. But there are practices of working on and engaging with and shaping the land worldwide, while there are many languages that do not have a direct translation for landscape: Arabic, Irish, Latin, and Yoruba, to name but a few.[79] As we will see from this book, landscape fieldwork can help us to understand those practices and nuances of language and meaning. Others, even the founders of the profession as we generally understand it, encourage searching for a new term for landscape architecture,

Figure 19. Günther Vogt, Fieldwork cupboard. (Photograph, VOGT Landscape Architects)

believing the compound to be too vague.[80] As I have argued elsewhere, perhaps landscape architects need to let go of the term "landscape" and embrace other terms from around the world: "green," "the architecture of particular places," "an engineer specialized in designing green spaces,"[81] to use concepts from around the world.[82]

For now, let us ask a more specific question: What is it about landscape that requires a particular form of fieldwork? Landscapes are spatial. Landscapes are grounded on the earth. They are also temporal, subject to the time of the day as well as the seasons and mature over decades and centuries. Landscapes are relational, depending on multiple assemblages of human and nonhuman relationships across space and time. To grasp these layers, I distinguish four qualities of landscape fieldwork. To do landscape fieldwork, you need to

1. Mix measures
2. Immerse your body
3. Contrast media
4. Critically imagine

The first speaks to the fact that landscape fieldwork involves some form of measurement, whether of space, time, or human values. Landscapes are complex and layered. Because of the complex relationalities of landscapes, they require a deeper form of engagement that mixes quantitative and qualitative measures. The second speaks to the in-depth nature of fieldwork. Understanding the physical, social, historical, and spiritual dimensions of a landscape requires a period of immersion in a field site through a process of what the anthropologist Subhadra Mitra Channa terms "embodied engagement."[83] The third highlights the need for multiple forms of media for documenting and interpreting a landscape. Lastly, landscape fieldwork involves an imaginative and projective dimension. Not all the above fields put this future-oriented dimension to the forefront. A landscape architect's purpose in doing fieldwork is distinct: we do fieldwork because we contend that by understanding a landscape, we are better equipped to intervene in it.

Five Forms of Landscape Fieldwork

In each of the subsequent chapters, I will discuss one of my own landscape fieldwork projects, each very site-specific. The chapters emphasize chance and serendipity, the use of visual information to represent ethnographic data, collaboration, and the application of anthropological skills to the study of landscape, materiality, and design processes. Each of the chapters, in its own way, uses "thick" ethnographic observation and description, applies theoretical concepts in making connections between ethnographic data, and moves toward landscape fieldwork as a distinct method for understanding how context informs design.

Based on the premise that the best way to learn how to do landscape fieldwork is simply to "do it," and the second-best way is to learn from what others have done, the book's chapters are punctuated with references to the landscape fieldwork of a range of designers and planners. These references introduce readers to fieldwork as a way of describing relationships between humans and humans, humans and nonhumans, or nonhumans and nonhumans. When we unravel the web of these relationships, we can design in ways that are ultimately more responsive and therefore more effective.

Chapter 1, "Designing at Home," builds on the prologue, where I introduced the landscape of the Irish / Northern Irish border and the agency of public space in peace and reconciliation. The chapter is an account of a modest public square that I designed and built in my hometown in the north of Ireland. This chapter traces the development of the project from concept to built form through a form of public engagement that ranges from residents to presidents. The project touched on many levels of hierarchy and demonstrates that landscape fieldwork is not just about observation, but also about doing. We'll see how landscape is profoundly political and how Irish understandings of landscape are markedly different to British concepts of landscape. This project sets up a conundrum about designing at home that I attempt to answer through other projects.

Chapter 2, "Throwing Beans," describes a form of landscape sampling, a seemingly random fieldwork method of throwing beans on a map and going to the points on which the beans fall. Celebrating chance and serendipity, the chapter focuses on a case in the Netherlands that I was engaged in when working with the London-based office CHORA/ Raoul Bunschoten. For CHORA, a landscape consists of "operational fields." These operational fields are the basic building blocks shaping the landscape and include everything from specific buildings to people's memories to economic trade routes. Throwing beans allows the landscape fieldworker to identify operational fields and use them as part of a multistep design process that includes developing prototypes, testing them through scenario games, and developing action plans.

Chapter 3, "Walking and Talking," reflects on a year of walking through Bahrain and conducting ethnographic research on the concept of "green," in the environmental sense of this word, in a country where water and greenery are limited and intertwined with greenery. I used Google Earth to determine the walks I would make through the Bahraini landscape, often at night. This fieldwork, conducted for my doctoral studies, was published in book form as *Paradoxes of Green: Landscapes of a City-State* (University of California Press, 2017). I walked in order to talk. I discuss the knowledge gained from unstructured interviews and chance conversations, as well as the value of an extended period in the field. I also discuss my frustration when I realized the limitations of ethnography to address urgent environmental issues within the Arabian Peninsula, and I reflect on the need to develop alternative means of landscape fieldwork that can move more quickly.

Chapter 4, "Engaging Collectively," focuses on an experiment in collaborative fieldwork in The Bahamas through a four-year research project. Instead of one fieldworker spending a year in a landscape, this project proposed that fifty-two people each spend a week in the place, and then combined those results. Would that research add up to something equivalent? As the chapter describes, while collaborative fieldwork did lead to rich data, the challenge became how to share and subsequently analyze the research. I discuss people-centered fieldwork as a complement to traditional methods of community engagement and public participation as an integral component of a design project.

Chapter 5, "Gathering Leaves," describes the process of doing a slower form of ethnography in Afro-Brazilian religious spaces, or *terreiros,* in Salvador da Bahia, Brazil. This method came about as I did not have time to conduct a continuous long-term study, but instead spread out my fieldwork over short periods conducted over several years. I ground this chapter in one Afro-Brazilian sacred grove, a place shaped by racial inequality and religious intolerance that, at the same time, offers remarkable ecosystem services to the city. Through the experiential knowledge gained from the field, I see these spaces are heavily designed, just with a different logic than is usually taught in Western-centered design schools. I also discuss the struggle I had over how much, if at all, to intervene as a designer in the space.

Chapter 6, "Thick Prescription," concludes the book with the reasons why landscape fieldwork should be about more than just observation and description, but about action and prescription too. I also reflect on what I see as the guiding ethos for the practice of landscape fieldwork, and where I hope to see the practice going in the future.

Reflections

In this chapter, I have introduced the way that the founders of the field of landscape architecture were formed by fieldwork. Using the example of the magical garden at Shute, I describe it from the perspectives of various fields. Through that landscape of landscape fieldwork, we see how it draws from many genealogies and the landscape fieldworker's tools may include the tools of the archaeologist, ecologist, geographer, and many others. Through landscape fieldwork, we find out some things about the garden that we might otherwise not have known. Landscape fieldwork is distinct for being spatial, and projective; and the landscape fieldworker is constantly imagining the future.

Landscape fieldwork allows us to challenge received norms, and the orthodoxy of our approaches from a disciplinary perspective. Relying solely on landscape architecture precedents that have been documented in written histories and often produced under rigid power structures privileges some voices, such as heterosexual cisgender white males, over others. If we as a field want to challenge, grow, develop, and diversify ideas and ways of working, landscape fieldwork holds the key. If you are interested in people, and the diverse ways that people interact with landscape, and how an understanding of both can help us be more reflexive designers and scholars, then I invite you to keep reading.

1

DESIGNING AT HOME

Building a Diamond in the Irish Northwest

There's a name for every stone about here, sir, and a story too.

—Brian Friel, *The Gentle Island*

I first sketched the plans for the Diamond over evening tea in the main restaurant of University College Dublin. It was 1993, and I was an undergraduate student of landscape horticulture, in the second year of a five-year program. The question in my mind was how to resolve the change in levels across the triangular town square in my hometown, let's call it Ballybeg, located centrally at the heart of the Inishowen peninsula in the north of Ireland, about four hours north of Dublin.[1] Although shaped like a triangle, this area at the town's center was called Diamond, as are many public squares in the north of Ireland, whether any other shape—square, rectangular, or even triangular.

The question of how to redesign the Diamond had long puzzled the local authorities. In the 1960s, the Donegal County Council built public toilets there. The structure was level with the Upper Street, and was entered from the Lower Street, creating a roof that served as a sort of public space. But the structure's roof was not the

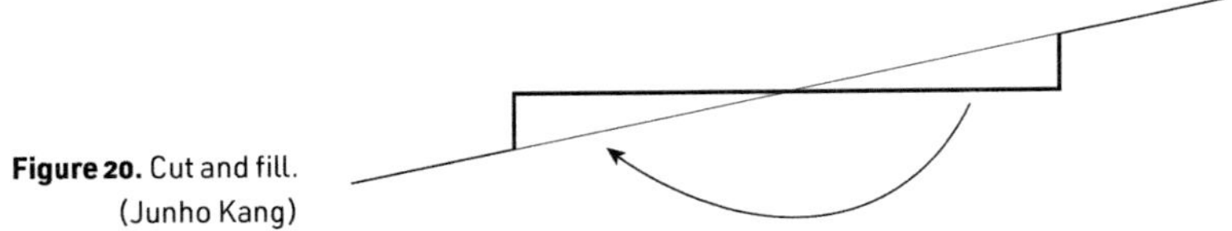

Figure 20. Cut and fill. (Junho Kang)

inviting space the engineer who designed it imagined it would be. While the roof was used for the popular annual Christmas tree festival, it was not utilized much for the rest of the year. I knew that I wanted to design something different—a well-conceived and functional space that met the needs of the community I knew so well.

As I doodled on the paper napkins while eating baked beans, eggs, and chips, I realized the three-meter (nine-foot) distance between the Upper and Lower streets could be resolved using the time-tested method of cut and fill. With cut and fill, the higher level is excavated and the material that is unearthed is used to raise the lower level. This strategy, I came to realize, would also mean that the median level of the public space I was proposing would connect with the outside street. Excited, I hurried back to the studio, working late into the night to translate the rough sketches from the napkins to more measured drawings. These first studies confirmed my hunch that cut and fill could work. This redesign of the Diamond was the first formal design project I had ever undertaken. As I sat at the drafting table, I felt a huge sense of anticipation. How could such a simple solution, straight out of a textbook, answer such a seemingly complex question that had perplexed people for years?[2]

Fieldwork Design

In this chapter, I will describe the process of designing and building the Diamond, one that I witnessed from the unique perspective of designer, local resident, client, and contractor. Usually, a project is experienced through only one of these lenses. Occasionally, a designer is also a contractor, or a local resident is also client, but to actively engage in all four perspectives is very rare. As I outline below, a peculiar set of events and a complex web of relationships put me in this position.

My involvement in the design process, likewise, was not linear; it was lengthy and continued right up to and even after construction. Over a period of months, I went through many versions of the design, though the basic proposal, based on cut and fill, remained the same. I questioned the proportions of the public space, its material composition and form, the number and location of parking spaces, the species and locations of the trees, and even the location for the Christmas tree. As I came to understand, the project also had many other dimensions that went far beyond its functionality. These related to the multiplicity of symbolic and interpretative dimensions of landscape architecture, which could only be understood through community engagement and participation.

A central question that designers should always ask is: "Who gives us the authority to design?" The answer to this question will be different for every project. In the Diamond's case, through deep social entanglements, the townspeople allowed me the authority to design for the town—but the process and its conclusion were far from straightforward. Though I did not describe my work then as emerging from, or as, "landscape fieldwork," this was in fact my first foray into using that method for a design project. The fieldwork process was distinct, however, as it engaged with opinion leaders and those in positions of public authority as much as any other resident. In my approach, I tried to challenge social hierarchies and engage a range of people from residents through to the president. Fieldworkers tend to work with one or the other social group—"top down" or "bottom up"—and this fieldwork managed to cut across social distinctions. It didn't start or stop with site analysis but engaged with the act of design and construction, and maintenance. And it began with understanding the history that shaped the place where I was working.

The Dilemma of the Diamond

The town or village is situated at the northern head of Glentogher in the center of the Inishowen peninsula (*Inis Eoghain,* the Island of Owen), the northernmost peninsula in Ireland, further north than Northern Ireland. The south of the peninsula is joined to the rest of the island of Ireland by a relatively narrow isthmus, formerly sea, which could be sea again sooner rather than later due to the impacts of rising sea levels because of climate change. Accessible mostly by passing over and through mountains and valleys, it's no wonder that the peninsula was one of the last parts of Ireland to be conquered by the English. The valley where my hometown is located is where St. Patrick, a legend has it, drove the last serpents from Ireland. Around thirteen miles (or twenty kilometers) long and three miles (five kilometers) wide, the parish is roughly the same area as Manhattan. Situated about two miles (three kilometers) from the sea, on a good day one can see on the horizon the Paps of Jura in the Inner Hebrides of Scotland. The proximity to Scotland is one of the reasons the local dialect is closer to Lowland Scots than the Hibernian English spoken by the rest of Ireland.

This is historically the land of the Ó Dochartaigh clan or, as the name has been Anglicized, the Dohertys. The name originated in the eighth century, so I can assume that my family have been in the area for over 1,300 years. There are so many

Figure 21. Detail from map of Ireland by cartographer Jan Jansson, ca. 1650. Published by Joan Blaeu, under the title "Vltonia; hibernis Cvi-Gvilly; anglis Vlster," in *Hibernia regnum: Spezialkarten: Sammelband der Ryhiner-Kartensammlung* 7, [Amsterdam]: [Joan Blaeu], [1654]. (Image from Universitätsbibliothek Bern, MUE Ryh 2003: 7, https://doi.org/10.3931/e-rara-109629)

Figure 22. Location. (Maria Vollas)

Dohertys that residents must use other nicknames to distinguish between families. These nicknames, which continuously unfold, sometime relate to a trait or trade—such as "the Mason" or "the Saddler"—or to a place—such as Liss or Umgall—or even to some personal or physical resemblance with an historic figure—such as John the Baptist, Napoleon Bonaparte, or Osama Bin Laden. They might also be a combination, such as my own family's nickname, "Fintan," derived either from a town called Fintona, or more likely an executioner called Fenton.[3]

In addition to people having more than one name, the land too has multiple names. Some come from English, others are transliterated from Irish Gaelic, and still others persist in Irish, such as the two hills that dominate the town: *Sliabh Sneacht,* or "Snow Mountain" (the highest nearby peak at about six hundred meters or two thousand feet), and *Cnoc na Coille Dara,* or "The Hill of the Oak Wood." Residents have a deep affection for the land, as immortalized in the work of local playwright Brian Friel. Brian's wife, Anne, went to school with my mother, and I first encountered Brian in my family's shop. Manus, a character in Friel's play *The Gentle Island,* tells us, "There's a name for every stone about here, sir, and a story too."[4] This attachment for the land is evident in the word *baile,* which can be used interchangeably in the Irish language for a home, a village, a town, or a city. For example, Dublin is *Baile Átha Cliath;* and a small Inishowen village, home of a famous golf club, is *Baile Lifín* or Ballyliffin. To say one is going home, one might say *Tá mé ag dhul abhaile,* with *abhaile* meaning home.

This linguistic inclusivity speaks to understandings of landscape that existed before the British conquered Ireland. The native Irish did not live in towns or villages. They mostly lived in dispersed settlements known as a *straidbhaile*—a "street town" whereby extended families built their houses on the "street," or yard, of the parents—or as a *clachan.*[5] As a result, the concept of towns, villages, and even cities did not exist. Instead, land was communally owned and farmed by clans. When the British colonized Irish lands in the early seventeenth century, in an effort known as the Plantation of Ulster, it brought an end to the Gaelic order. With the flight of the traditional leaders to the European mainland, the English set about building a network of towns and villages.[6] An objective of these settlements was to establish control over the landscape through providing spaces for commerce and the privatization of property.

Settlements dating from this period have one feature in common: a central market square called a Diamond. In my hometown, the triangular Diamond, which sits at the intersection of four main roads with two roads converging on one of the angles, had been the site of a marketplace ever since the town was founded in the 1600s. The ruins of the previous settlement exist to this day on a nearby farm overgrown with hawthorn (*Crataegus monogyna*) and whins (*Ulex europaeus*). Markets were held weekly in the Diamond, always on Mondays and sometimes, during busy seasons, on Thursdays. Fairs were held quarterly. The main activity of the markets was the exchange of farm animals, mostly cattle and sheep, but it was also where "kenters" came to sell clothing and household goods, usually traveling from twenty miles (thirty-two kilometers) north from nearby Derry-Londonderry.[7] Servant boys and girls could also be engaged at these markets until the mid-twentieth century. My grandparents were "hired out" as servants at the age of twelve, and this was common in the area at the time.

The commercial activity was facilitated by a market house, which was used by officials for registering sales and gathering taxes. In 1867, a local historian writing under the pseudonym Maghtochair, described it as, "a large structure, of rather ungainly aspect, serving the purpose of a weigh-house and Sessions House but impairing considerably the general appearance of the town."[8] The market house had an upstairs room facetiously called the Crit, named after the Criterion Ballroom in London's Piccadilly, which served as a space for social events and weekend dances. The glamour of central London was, seemingly, a far cry from a market house in the periphery of rural Ireland—but the name was also a marker of the British Empire; the area belonged to the United Kingdom until 1922, and the "Crit" moniker persisted after independence. In the 1950s, the market house was demolished due to its delipidated condition, and the market was moved to a newly built indoor facility. At the time, people told me, they were happy to see the market house demolished. Its demise opened up the town by creating views across the Diamond. Nevertheless, the change in levels left behind where the market house stood made the Diamond too steep to be used on an everyday basis.

As mentioned, in the early 1960s, the local government, Donegal County Council, built a block of public toilets where the market house had been. The logic of providing public conveniences in a village of only two hundred houses and a population of about three thousand people might at first seem questionable. But at that moment in the 1960s, it was surely seen as a sign of progress. Just as the cattle- and sheep-dung-covered streets were asphalted and sanitized during that period, public conveniences like toilets were constructed to wash away human waste. Moreover, on market days on Mondays and Thursdays, when farmers and kenters came from across the peninsula, it easily doubled the population in town. Thus, the provision of public conveniences was seen to be addressing a public need. Shortly after the toilets were built, the market itself was moved out of the Diamond, to the purpose-built, corrugated-iron-covered structure called "the Mart," meaning market, located just outside the town. But the public toilets remained. As did a weighbridge for farmers to weigh their produce, and a fire siren—used to assemble the fire brigade in the predigital age.

I grew up in the Diamond in the corner shop, a space that many community members encountered daily when they picked up their newspapers and other sundry items. To the rear was an ice cream parlor. Before my father left this world, when I was just thirteen months old, my parents ran a restaurant, which was busiest on market days. After the restaurant was closed, that room became the "pool room," used for playing pool and table soccer, or fussball. The shop acted as a space for community members to linger and talk as they bought their newspapers, and many couples in the community met either in the ice cream parlor or pool room. It was, to quote the poet Seamus Heaney, "The rooms where we come to consciousness."[9] From an early age, I came to know the townspeople, and they me. If they did not know me, the youngest in the family, for sure they knew my mother, my late father, or one of my sisters and brothers. Small-town life can be much more complicated than in a city due to the multitude of unavoidable social relationships, many of them inherited.

From my point of view as someone intimately connected with the community, the new public toilets dominated the town. I grew up looking at the toilet structure across the street and wishing they had never been built. My home on the Lower Street faced the ladies' toilets while my grandfather's house across the street faced directly onto the men's toilets. On the few occasions I ventured in there, I was scared by the cold and dampness, never mind the smell. No wonder, therefore, that I decided to pursue the redesign of this space as an independent study years later when I was at university. On the surface, the design project seemed to simply be about remaking a public town square to be more functional. But it turned out to be more complicated because of the social, political, and economic context. At the time, Ireland and Northern Ireland were going through a peace and reconciliation process, which brought up long-standing tensions in the community. These conflicts emerged in every aspect of daily life, including in negotiations over the design of the Diamond.

First Sketches

The initial sketches did not take long to draw up. I drew the Diamond using the conventions of plan, section, and elevation, as I was mainly concerned with grading of the street levels and in ensuring the scheme would function practically. While I did not at first have a formal survey of the streets with the spot levels to work from, I knew the Diamond well enough to make informed guesses of the levels. After all, I had lived there all my life and played on the streets as a child. I was deeply familiar with the peculiar nuances of the structure and the surface materials of local pebble dash and concrete, and I translated that to drawings. Until I conducted a formal survey, for the purposes of the sketch design, I took the rooflines of the buildings that fronted the Diamond as my base level and derived all my measurements from them. It helped that I also knew the people who lived inside the houses; I knew where they were likely to fit into the economic, political, religious, and sociocultural landscapes, which also have their physical dimensions. In other words, being from the town gave me access to the aesthetic values and limits of the residents.

Precedents: The Donagh High Cross

The cut-and-fill plaza made sense to me as a technical solution, but what should the plaza be? What form should it take? Should the center plaza be triangular, echoing the outside street? Square? Circular? Oblong? Amoeba-shaped? A diamond? To answer these questions, I began to look for inspiration in local symbols and artefacts. The former monastic settlement is the location of the famous Donagh High Cross. The noted art historian Françoise Henry dates the cross to the seventh century. Henry tells us that it was the first time the cross broke free of the slab.[10] Crosses had previously been carved onto the face of a rectangular stone; this was the first time a stone itself was carved by stonemasons into the shape of the cross. Henry writes: "There was a dramatic moment when it looked as if some driving force was struggling to shape the inanimate mater. The idea of the cross was still embedded in the stone, unable to free itself, yet pushing out at the sides and the top. Finally it fought its way out of the inert frame and the first free-standing crosses appeared."[11] The cross has several notable features. For one thing, it lacks symmetry. Henry writes that it was carved by someone who "shrank from the use of straight lines."[12] This asymmetry could be because of the resistance of the stone to the stonemasons' chisels, although sandstone is relatively soft. More likely, it could be due to a *laissez-faire* attitude of the craftsmen who on carving the stone may have said, "Ah, sure, it'll

Figure 23. The Donagh High Cross. (Photograph by David Lyons)

do!"[13] Another key feature of this cross is its orientation toward the east. Most Irish high crosses of the period were oriented to the west, as a sign of penance, because the setting sun symbolized the death of Christ. The front of this high cross, on the other hand, looks toward the rising sun, which can be symbolic of new life and growth. No one knows for sure if the cross always faced east or if it was moved at some point and turned around.

The interlacing pattern on the cross is also highly distinctive and is found in only three places in early Irish art, the other two being St. Mura's Cross in the nearby hamlet of Fahan, and the *Book of Durrow,* an illuminated manuscript from the same period. For this reason, the interlacing pattern is found on all sorts of branding associated with the town. It's the logo for the local high school, where I went to school, for instance. The interlacing pattern also appeals across the community. As a symbol that existed before the Reformation and before the English invasions of Ireland, even before the Vikings arrived, the symbol was as apolitical as any symbol could be in Ireland.

Paving Pattern and Enclosure

The solution I settled on was to reuse this ancient interlacing pattern as the paving on the floor of the plaza. The paving would be constructed using two bands of contrasting colors of locally sourced stone, highlighting the entwined nature of the bands as well as the materiality and colors of the surrounding landscape. The paving pattern would be seen from the upper street as the visitor looked down into the space, and it would reveal itself to the visitor entering from the lower street, as well as from the mid points. It would also be clearly visible from the upstairs of the

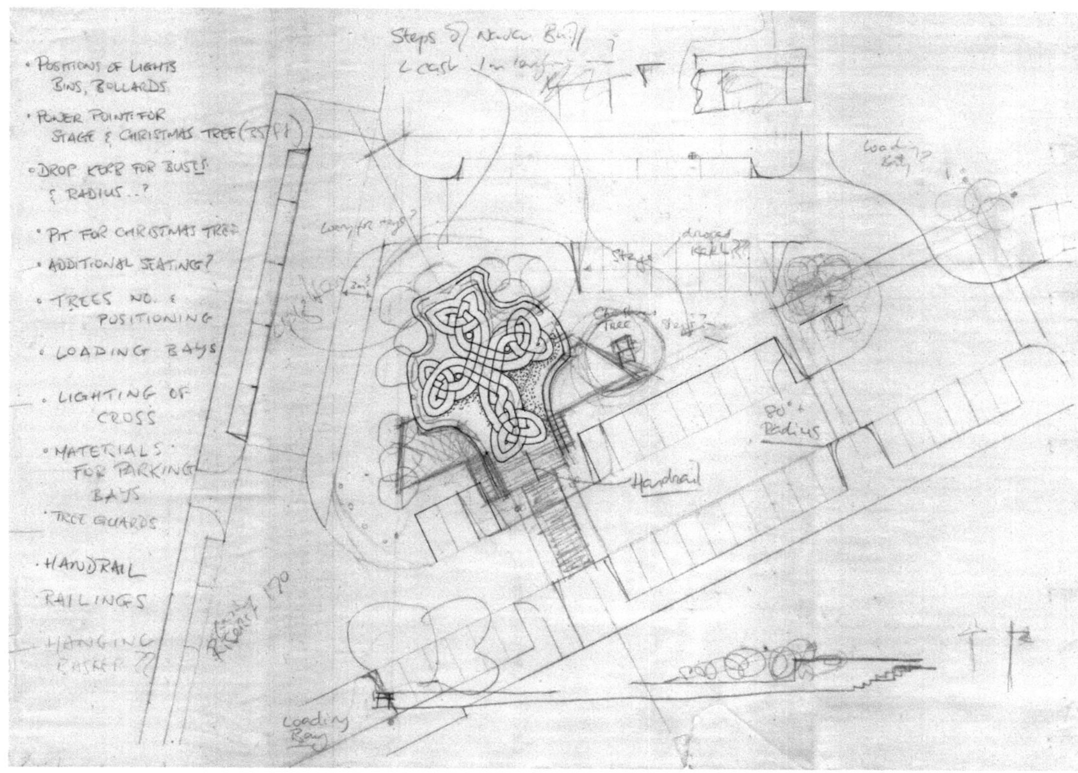

Figure 24. The Diamond plan, initial sketches, ca. 1993.

buildings that lined the Diamond. While I sketched many versions of the pattern applied to the site, readjusted to fit the proportions of the plaza, the scale that I eventually settled on would be to enlarge it to between sixteen or seventeen times the size of the original.

In some ways the pattern would function as a combination of Robert Venturi, Denise Scott Brown, and Steven Izenour's concept of the "duck" versus the "decorated shed." A duck, they explain, is "where the architectural systems of space, structure, and program are submerged and distorted by an overall symbolic form."[14] By contrast, the decorated shed could be seen as a sign or billboard crying out what makes it distinctive in some way.[15] Their work on Las Vegas was based on fieldwork carried out during a studio at the Yale School of Architecture. (I was later to study with Steven Izenour and work as an intern in Venturi Scott Brown's office just outside Philadelphia.)

I was also motivated by the work of the Brazilian landscape architect Roberto Burle Marx, whose work I first encountered on a BBC television program.[16] Burle Marx, as I noted in the introduction, was a flamboyant modernist known for his abstract paving patterns, and use of native Brazilian plants. I was especially moved by Burle Marx's project for the headquarters of the Safra Bank on the Avenida Paulista in São Paulo. I studied the images in depth: the juxtaposition of colors, materials, textures, volumes, and especially the lines.[17] Burle Marx's curves were distinctive, inspired by aerial views of the Amazon River and other rivers that run through the Brazilian landscape, which he witnessed from airplanes during his regular field expeditions to collect plants.[18] In addition to his abstract paving patterns and artistry as a designer, I was also deeply affected by his lecture notes, which were given to me by his collaborator and heir, Haruyoshi Ono. In them, Burle Marx calls for landscape architects to radically change cities and society.[19]

Once I had settled on the paving pattern, I needed to figure out the enclosure for the Diamond, the physical separation from the outside street. My instincts told me that the central plaza should be surrounded by a wall of just over waist height, designed to give a feeling of separation from the street, but not too removed from it either. The wall would be set inside a grove of trees at the lower end of the plaza, the volume of the trees completing the feeling of enclosure. As it turned out, it was these proportions that proved the hardest aspect of the project to reconcile; I struggled

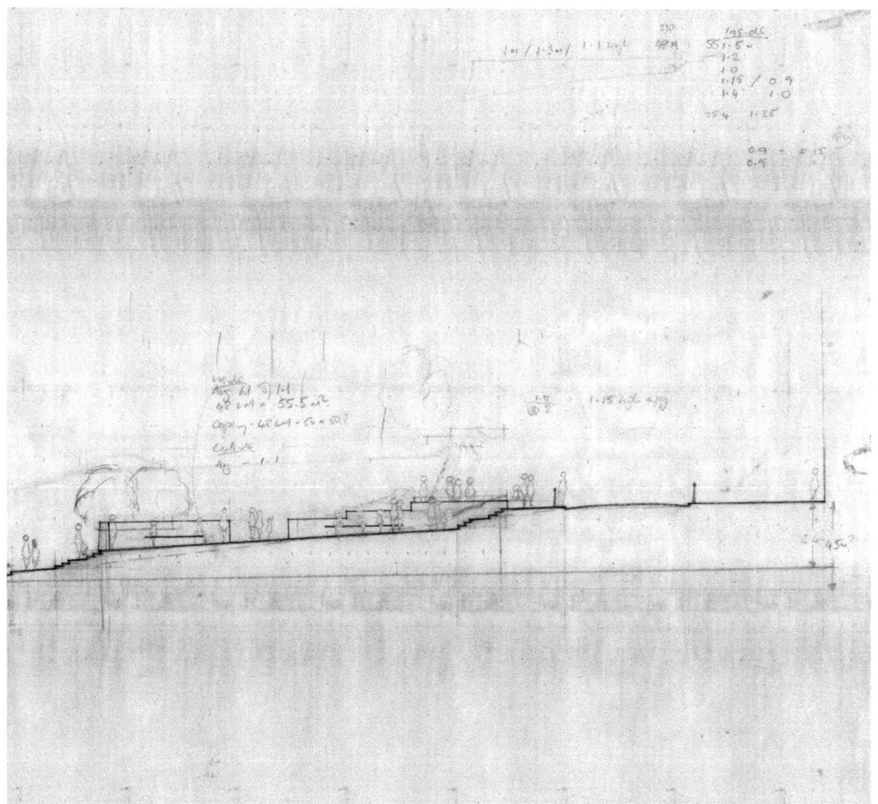

Figure 25. The Diamond section, initial sketches, ca. 1993.

with the dimensions of the pattern on the ground as well as the height of the walls, especially considering the slope.

As the plans developed, I realized that certain aspects needed to be sorted out in person as we went along. That's because the architectural conventions of plan, section, and elevation could not adequately capture the feelings of being in the space. I made models in three dimensions, which were more helpful than the drawings, but the actual details had to be worked out directly with the stonemasons during the building process. Likewise, I ended up choosing the trees that enclosed the walls in person during a visit to a tree nursery outside Belfast across the border in Northern Ireland. In other words, the design process was not linear. It involved lateral thinking and continued during and even after the project's construction.

Christmas Tree

As I developed the design, thinking through sketching, drawing, doodling, I would return home at weekends and discuss my ideas with my family and neighbors. The main question on their minds was: "What will happen to the Christmas tree?" My hometown then sought to have the biggest Christmas tree in Ireland every year. The Christmas Tree Group, made up of volunteer townspeople, would scout for trees in the commercial forests in the surrounding hills and countryside. The tree would be erected on the Sunday before December 8, the Feast of the Immaculate Conception, which marked the start of the Christmas season. The tree was covered in multicolored fairy lights, then remained in place till the Feast of the Epiphany on January 6. In addition to these Christmas festivities, New Year's Eve was a major celebration that took place in the Diamond. The New Year was rung in by a dignitary, such as John the Saddler, an auctioneer who was directly elected mayor, a symbolic role. The ritual of ringing in the New Year dated back only to after the construction of the public toilets in the 1960s, but it had become very popular across the peninsula. But these annual winter events, in my view, were also evidence of the failure of the space; they were the only activities that people could imagine happening there.

I imagined this public space as something different: the space would still be used for the Christmas tree, which in many respects served a similar function to Nelson's Column in Trafalgar Square or the Diamond War Memorial in nearby Derry-Londonderry; it acted as a plinth, obelisk, and column all in one due to its vertical dimension. But, more importantly, it would be a plaza that could be used and enjoyed daily and year-round by townspeople. To this end, there would be a seating area in the center of the plaza surrounded by a wall. But, I imagined, the space could be used as much by being looked at as actually being inside it. Even if no one sat inside, the plaza would contribute to the town by providing an aesthetic and social use.

Plus, I set aside parking spaces for the market traders in my proposal. Although the cattle and sheep trading had shifted to the mart in the 1960s, I knew this was still a market town. Traders would set up their stalls on Mondays, following from the pattern of the market days, and sometimes on Saturdays too. On Fridays, the "fishman," whose real name I never knew, would come all the way from the port of Killybegs about ninety miles (144 kilometers) away. (I could tell he was a native Irish speaker because his signs would read "mackerl" and "sammon" rather than mackerel and salmon.) I knew parking was important to people.

According to one of the oldest residents in the town at that time, Ellen McIntyre, "Where are you loused?" (meaning, where did you leave your horse?) was a common question every market day. The standard reply would be "Up in George's back," meaning the George the Saddler's backyard. Clearly, parking was an important topic even before cars.

Building Allegiances

Once I had a plan sketched out, I presented it to my professors at University College Dublin and received their critique and advice. My professor and project supervisor at UCD, Milia Tsaoussis-Maddock, was originally from Greece and had studied landscape architecture under Ian L. McHarg at the University of Pennsylvania. Milia later inspired me to study at Penn too. Milia offered detailed suggestions on the design and its representation through drawing. After adapting the design details accordingly, and receiving their approval, and a grade, the next essential step was to present the set of drawings in the community itself.

In addition to continuing to discuss the plans with family and friends, I next talked with the local development association. This was a group of volunteer townspeople, mostly businesspeople, who had assumed similar roles to the local village government. They had little budget, and they exerted power by lobbying for the improvement of the built environment. Some residents claimed that the development association was a self-interested group that had no authority to act on behalf of the townspeople. Still, for my purposes, they offered a necessary bridge to the local government. The development association, after looking at my design, told me that they liked the plan very much. They liked the fact that the Christmas tree would be accommodated in the center of the plaza, and especially the organizing of traffic and parking spaces. This, I was later to see, was a bone of contention for the development association. Many of them wanted the street traders to be eliminated from the Diamond. They used the argument that there were not enough parking spaces as the reason, but in private the real reasons came out: they were frustrated that the market traders did not have to pay the commercial rates that formal businesses were required to pay. They wanted to eliminate their competition from the town.

In hindsight, I can say that the design extended beyond drawing and the "fieldwork" extended beyond the immediate physical landscape. I had to learn how to navigate local political alliances and social tensions that were very specific to this place. Along the way, I realized that my familiarity with the town and its residents is what helped me to succeed where others might not have. Fieldwork, as active engagement with a site and its users, was central to the project getting built. My fieldwork brought me wide and far, not just within the community, but beyond. Most notably, in the process of getting advice on my project, I reached out to the well-known landscape architect Sir Geoffrey Jellicoe, the designer of the garden at Shute.

Disciplinary Approval

While working on the design for the Diamond, I remembered that I had seen a photograph of the Donagh Cross in volume 2 of Geoffrey Jellicoe's *Studies in Landscape Design.*[20] This series of three volumes were my favorites of Jellicoe's works. As I reread the book, I realized Jellicoe had been to my hometown, and the photo was taken in the 1960s. On this basis, I decided to write him a letter. Sir Geoffrey replied right away: "Give me a ring and come and have coffee at eleven and a chat of one to one and a half hours." I gave him a call, as invited, and we arranged a date in January for me to visit him at his home in Highpoint, in Highgate, north London.

Jellicoe's home was designed by Berthold Lubetkin (1901–1990), a Russian-born architect famous for his contributions to British modernism. The modest apartment, notable for its straight lines, was decorated with modernist paintings that Jellicoe said filled him with, as he put it, "electricity." The artworks were mostly on one wall that he called his "wall of fame." Jellicoe told me that he didn't leave the house anymore, but instead the world came to him. From his flat on the fourth floor, we were greeted with sweeping views over Hampstead Heath to the north—a disparate urban landscape held together by trees. On the fourth floor, one is still connected to the trees: any higher and they would be too far below, Jellicoe told me. Jellicoe invited me to stand on the balcony that led directly from his living room. Standing on the space—just big enough for one person—I lost awareness of the railing and had the impression that I was floating above the landscape below and in front. As Sir Geoffrey remarked, that was the point! It was hard to believe we were so close to the center of the capital.

I had gone to Sir Geoffrey with a list of interview questions mostly relating to his and Burle Marx's design processes. I was working on a thesis arguing they were extremes in modernist approaches to landscape architecture. But Sir Geoffrey first asked to see my drawings of the Diamond. Later, he explained why. He said that from seeing my drawings, he could see how I think. As I unrolled my drawings, with the cross's interlacing pattern reproduced on the ground, I explained to Sir Geoffrey that I had an inner crisis with the design. I was in turmoil. Was I copying from the past rather than looking to the future? I quoted the artist Paul Klee, "Art does not reproduce the visible; rather it makes visible."[21] I quoted Sir Geoffrey himself, "Let us by all means and at all times weigh our endeavours against those of history but let us never accept the past as final and become copyists."[22]

Sitting on the floor beside Sir Geoffrey, who was seated on an armchair, we discussed my plans for the Diamond. At first, looking at my sequence of drawings, he was coy. He said he admired the placing of the grand idea of the cross in the center of the square, which he saw as tapping into the subconscious of the local people. I showed him a scheme with a seating area inside, and that was the one he liked most. He approved of how the central plaza was a different space relative to the town outside; it should be in a world of its own. He went on to tell me, "You're attempting one of the boldest things ever a landscape architect could do: set a very grand idea into a shape which is a commonplace shape." But he felt the proportions of the design weren't right and told me that was something I needed to work at. He also did not like my drawings; I needed to work at my colors and at the luminosity of my renderings, he insisted. Still, in the end, Jellicoe reassured me that I had the right balance between the interpretation of the past and the need to be projective. "That job you are doing are for those people there, it's not an individual I don't think, or anything like that, it's for the people," he reassured me. "And I think you got the right answer."[23]

This was the first of four formal interviews with Sir Geoffrey that I conducted in 1994 and 1995. During those sessions, and through our correspondence, I kept Sir Geoffrey abreast of the unfolding developments with the project. Later, in April 1995, Sir Geoffrey wrote a letter officially lending his support to the project. He wrote: "Although too long ago for me to remember the site of your proposed garden, I consider the idea of it to be practical and most beautiful. When completed it would manifest the poetry and nobility of Ireland and soothe and uplift the spirits of those who rest in it. Yours very sincerely, Geoffrey Jellicoe." Jellicoe wrote the letter to add it to the public exhibition, which was held later. I liked the fact he called it a garden rather than a plaza. The idea of having a garden in this marketplace was at first antithetical but later emerged as a key part of the concept.

What was essential from this interaction with Sir Geoffrey from a fieldwork perspective was the engagement with the elites, opinion leaders, decision makers, and influencers, as a complement to fieldwork within the community itself. Too often fieldwork is assumed to be a protracted engagement with so-called ordinary people—which is essential—but it is important to recognize and respect the position and agency of the designer. In this case, Jellicoe's support grounded the project more deeply in the field of landscape architecture, a discipline that few locals understood. In fact, the Northwest at that time had scarcely any landscape architects. People did not understand what landscape architects did, but they knew it was something to do with design of the exterior, and usually something associated with large country houses and estates. I was very aware of my own role in the town as a son, grandson, and brother. How would the townspeople ever trust me, a "green" student, with the redesign of the Diamond? Having Sir Geoffrey's support added a layer of professionalism to the endeavor that proved important in securing community support.

As I left Sir Geoffrey's apartment building after the interview, I had another encounter that took me out of the immediate issues with the design of the Diamond and ended up staying with me throughout the process. I went around the rear of the building to take a photograph looking up to Sir Geoffrey's flat and the balcony I had stood on. As I pointed my camera, I was stopped by one of the residents to inquire what I was doing. When they heard my accent, they said they had to be very careful with Irish people, like me, in London. The IRA (Irish Republican Army) had detonated bombs in Bishopsgate in the City of London the previous year. This brought home to me the link between the Irish struggle with the center of colonial capitalist power, and how these forces come together in a small village square.

Gaining Support

On securing the support of Sir Geoffrey Jellicoe, and the development association, the next step was to seek the approval of the Donegal County Council, the local government. The Council predates the Irish independence of 1922, having been established under the foundational Local Government of Ireland Act of 1898. Each area has several councilors who either represent various political parties or sit as independents. I presented the plans to several council members, all of whom showed their enthusiastic support for my project. I had expected to be rebuffed or at least to be told the plans were unrealistic, so I was genuinely surprised by the positivity of the reactions that I encountered.

I also met with the council's local administrator, Richard Flynn, who was based in the town and from the town, too. At first, Richard, a no-nonsense engineer, intimidated me with his gruff manner, but he saw value in the scheme and became one of the project's biggest supporters. I discovered through working closely with Richard in the months and years that followed that he had a big heart. He introduced me to Faye Cunningham, the chief planner for the county. Faye headed the Forward Planning Unit, and all projects such as this one would come through her office. Faye had a fascination for Japan and had traveled there several times. Faye told me that one of her favorite books was written by an anthropologist, Ruth Benedict. She explained to me that *The Chrysanthemum and the Sword: Patterns of Japanese Culture* describes the seemingly contradictory sides of Japanese culture: the aesthetic and gentle side of the chrysanthemum combined with a ruthlessness that comes from wielding the sword.[24] Faye herself was a gentle soul, but she could be firm-handed too. I wondered if Faye admired this perceived harshness of Japan and tried to apply it in her work with the Council.

In due course, in March 1995, the plans were officially approved by the County Council as their official proposal for the town. But, even then, I don't think anyone really expected it would be built. I suspected they supported it because they wanted to see change, but to my knowledge there was no realistic expectation that public funds would be available to build it. In this, gaining the support of two public figures, Bernard McGuinness and Denis McGonagle, proved to be pivotal. Bernard was chairman of Donegal County Council when the plans were first drawn up and when it was adopted by the council. Denis McGonagle, from the village, was chairman when the project was opened. Another key individual was Michael McLoone, the county manager. The determination of these three locals was critical in ensuring the project received the attention of the county authorities.

Material Investigations

Another design doyen I consulted on the Diamond was Liam McCormick, a brilliant Irish architect who avoided the architectural scene and preferred to lead a quiet life in the Northwest, together with his wife, Joy, and children, Aisling and Finn. I reached out to Liam for advice, and we ended up becoming familiar with each other, often visiting each other's homes. I trusted Liam. His work designing churches stood out to me as one of the best examples of the architecture of the Irish landscape. Of the twenty-seven churches that he designed over his lifetime, the most well-known was the Church of Saint Aengus at Burt, not too far from my home, which won a gold medal from the Royal Institute of the Architects of Ireland in 1967 and in 1999 was voted "Building of the Century" in a poll in the Irish weekend newspaper, the *Sunday Tribune*. Liam also received a papal knighthood, Knight of Saint Gregory, for this ecclesiastical work.

Most importantly for my project, I was compelled by how the structures Liam designed were each deeply embedded in their landscape, taking their form from the landscape as well as materials.[25] In my very first conversation with Liam, I told him I had heard in one of my undergraduate landscape architecture classes that he had camped on the site of Saint Aengus for three days to experience the *genius loci* before embarking on designing the church.[26] He denied this, although he approved of the aspiration behind the rumor. He confided in me that the first sketches of the asymmetrical roof were sketched on a beer mat. I was struck by Liam's insistence that he thought of himself as a master craftsman, bringing together various artists and craftspeople to design a formal structure, as opposed to the architect being the sole design genius. In his acknowledgments, etched in glass on the door to Saint

Figure 26. Church of St. Aengus, Burt, Co. Donegal, designed by Liam McCormick. (Photograph by Henk Snoek, courtesy of RIBA Collections, London)

Aengus's chapel, Liam credited the builder and the craftsmen. Very few designers do that today.

An essential task that Liam helped me with was working out how best to craft the interlacing pattern on the ground from stone with a precision that the stonemasons of the original Donagh Cross had lacked. I was insisting on using local materials for the Diamond, but Liam suggested I use the "best materials," even if they weren't strictly local. He encouraged me to spend time visiting quarries and meeting stonemasons. Only after educating myself on what stone was available and who could build it and exhausting all the possibilities could an informed decision be made. Liam told me that in his experience there was probably only one firm in Ireland with the technical abilities to meet the ambitious task I had set. They were called Stone

Developments and based in County Wicklow, south of Dublin, a good five- or six-hours' drive away. I visited Stone Developments and met with the manager, Sean Potter. Although the stones they had available were all from County Wicklow, a seemingly long way away, their stones were of shades that were livelier than the dull grey-blues of the limestone from the local quarry that I had been insisting on using until then. Liam had been right.

Liam also helpfully put me in touch with Eunan Doyle, a creative quantity surveyor, who costed the initial scheme *pro bono.* Liam helped me understand that realistically costing a project was an essential aspect of the design—and doing so would make it more appealing to the County Council. The quantity surveyor was to be a friend, rather than an enemy. Eunan, he explained, was sympathetic to good design. I later learned that Eunan had a friend living in Lubetkin's Highpoint, although they did not know Jellicoe. In addition to Liam, and Eunan, I was indebted to John Reid of Crawford Campbell Associates, then one of the few landscape architects in Northern Ireland. John, who had taken over from Crawford Campbell, a Penn alum, was familiar with the mysterious workings of the council. John was always gracious and kind with his time.

Later, at the public launch of the plans for the Diamond, in May 1995, Liam spoke in public praising the project. He said that if he could choose someone to go into partnership with, it would be me. If Jellicoe's imprimatur was not enough—and there were a minority that saw the London-based Jellicoe as a foreigner and a titled imperialist—it was important the town's people heard this from one of their own. Liam was a well-respected retired architect, and his opinion mattered. Combined with Jellicoe's letter, Liam's comments helped the townspeople see design value in the project and assuaged those who doubted the abilities of a student from the town.

Inside/Outside

Having selected the materials and costed the scheme, as I got further along in the design process, I became increasingly aware that my own positionality within the town might be a hindrance—especially when I realized I might need the support of the national government in addition to the local government. Even though I was seen as a "local," as a landscape fieldworker I had to somewhat remove myself from that position too and appear more "professional." Many anthropologists who study their own societies discuss this tension of being an insider/outside.[27] Lila Abu-Lughod calls these individuals "halfies."[28] In this case, I needed to find a way to navigate long-standing political divides in the community that went right back to the Irish Civil War of 1922–23. Essentially, there were those who supported the treaty that established an independent Ireland, represented by the Fine Gael political party, and by those who did not support the treaty, mostly represented by the opposing party of Fianna Fáil.

Throughout the design process, I shared an apartment on Lower Mount Street in the heart of Georgian Dublin with other landscape architecture students from the same county, an architectural technician, and a Sierra Leonean who, under refugee status, was studying law and learning to speak Irish. The flat was just a few minutes' walk from government buildings in Merrion Square. I took advantage of the location to visit Dáil Éireann, the parliament of Ireland, in Leinster House. Members of parliament are elected through proportional representation for a multiseat constituency. The then constituency of Donegal Northeast had three elected members, one each from the main political parties, and one independent member from a local party.[29]

I was acutely aware of the sensitivity of local politics. If I showed the plans to one politician, I needed to show them to them all. I met first with Paddy Harte TD (*Teachta Dála,* member of parliament), our local Fine Gael representative. People would have expected me to meet with Paddy first, as my family were perceived as Fine Gael, my father's family having supported the treaty with the United Kingdom during the Irish Civil War. I corresponded with Paddy and talked with him on the phone. As far as I could tell, he liked the proposal. On another occasion, I visited the Dáil for an evening tour with the Young Fine Gael student group from University College Dublin. After a tour of the chamber, we adjourned to the Dáil bar. There, I met another local TD. I had the plans folded up in my breast pocket, so I pulled them out and showed them to him as he returned from the bathroom. He said, "I can't talk about this right now, but give my secretary a ring and come in and have lunch one day." That I did, and we met for lunch at the Members' Restaurant. He said he liked the plans very much and asked, "What would you like me to do about it?" To which I replied that I simply wanted him to be aware of the plans. Having secured the tacit support of two parliamentarians, I then needed to secure the support of the third, which I did through an exchange of letters.

The three layers of government, local, regional, and national were complemented by two further dimensions: at the level of community and at the level of the continent. I needed to secure support at these levels as well. My goal was to secure the commitment of the government to build the project. I had no idea then how funding would be forthcoming, or from where, but I knew it was important to have political support.

Community Engagement and Communication

To engage the community, the development association proposed holding an exhibition about the plans for the redesign. The exhibition was held in the Diamond in my family's corner store. Following the hours of my family business, it was open to all and sundry seven days a week from early morning till late in the evening, for almost a full year. The exhibition was financed by a modest grant from the local development group with funding from the European Union's LEADER program, which supported community-based and rural development.[30]

The exhibition was hosted in my large workroom—a former dining room from the days of my family's restaurant with tiled flooring and paneled walls. Those who had trouble reading project drawing plans had a model to help them understand the scheme. The model was the centerpiece of the exhibition, set up in the middle of the room. I had the model photographed professionally and compared these photographs of the model with actual photos from the street. These before and after shots—not long before Photoshop became the norm—were the most effective means of explaining the project. Additional drawings mounted on the walls showed the plan, section, elevations, and a narrative text. Most importantly, the exhibition allowed me as the designer to interact and engage with townspeople in a meaningful way. The window blinds were kept open, so those walking by could investigate it from the street. When someone would come to view the plans, we could look out the same window or else walk outside to the site to discuss the impacts of the proposed scheme on the town. Sometimes, my mother would make them a cup of tea.

The exhibition attracted a steady stream of townspeople, with especially busy periods after church services. Most were slightly amused by my set-up. Some viewers came out of curiosity but had no real opinion. Others had seen the plans in the newspaper and came with strong preconceived opinions. Sometimes it was hard to tell what they thought. While almost everyone offered words of encouragement, I was struck by what I sensed as an overall lack of belief that any change could really be affected in the Diamond. Often people saw themselves as so far removed from Dublin, that the state didn't really care about them. But it was almost as if people didn't believe they deserved design. There was a deeply ingrained pessimism that came both from the neglect that comes from coming from the periphery, a long-neglected cross-border area in Ireland, but also the lack of exposure to anything special. One person told me that I was suffering from "delusions of grandeur." I felt judged for being out of my station and for daring to imagine other possibilities.

Others arrived with worries. Their concerns were usually about one of three things: the Christmas tree, parking spaces, and the public toilets. Often, people had helpful suggestions about the various rights-of-ways in the town and about the parking spaces. One of the concerned residents, for instance, was John Joe the Contractor. Together with his brother Henry, John Joe was partner in McGonagle's hardware store ("the Contractors") that abutted one of the corners of the Diamond. Since many of their deliveries involved trucks and lorries, the Contractors' business had parking needs. Sitting with John Joe and discussing these needs one to one was very helpful and constructive. Not only did it help me to understand the Contractors' issues better, but it helped us both to propose a creative and acceptable solution that involved them sharing an entrance with the church.

Martha Freed played the role of a dowager duchess in the town. Living alone in a ramshackle building that had once served as a department store that had long ago gone into decline, Martha was difficult to please. But I was relieved that Martha was charmed by the elm tree that was to be planted outside her dilapidated shop and home; she told me she could see the potential for the public domain to help rejuvenate private space. Another local resident, Margaret Butler, and the local Inishowen Environmental Group wanted there to be even more trees in the plan. Patrick the Jeweler was thrilled by the way the project opened up views to his shop, which was previously blocked off; so was Jim Melly, at the opposite side of the Diamond. Jim and his brother Bernard, the local undertakers, or morticians, wanted to ensure access for their hearse. Bernard Melly was concerned about providing a sheltered space for what he called the "*lachicos*" coming home from nightclubs, an Irish term for "fools."[31] Richard the Sailor was worried about the boy-racers—youngsters driving at high speeds at nighttime that plagued the town for years—so I proposed a series of ramps to deal with that. Others were concerned at having a streetlamp situated outside their house and how it might attract "undesirables." By and large, there was no great fuss, and most of the concerns could be easily addressed and accommodated. Usually, what people seemed to appreciate most was being listened to. Providing that forum for listening was essential.

I did not rely on the exhibition alone as a source of public information. The project was referenced in the various local media and church notices. Father Morris told the congregation in the Church of the Sacred Heart to make sure they went to see the great plans. The Reverend Lappin likewise recommended it at the Hillhead Presbyterian Church. DJs on the local radio station, NWCR, encouraged residents to see the plans. Tom Farren, the village reporter of the biweekly local newspaper, *Derry Journal,* regularly featured the project in his columns, and the Boy Scouts included it in their weekly newsletter.

Meanwhile, the most important referencing came in the conversations that took place in homes across the community of three thousand or so people. At various events, including wakes and funerals, people would ask one another whether they had seen the plans for the Diamond or not. On average, a person dies every ten days in my hometown, and wakes (what we call "corpse houses") and funerals are to this day important gathering moments in the community. Word of mouth was an important way to build knowledge and interest and to bring the plans into consciousness.

Once, the prime minister, Taoiseach John Bruton, called to see the plans. He was brought there by the county chairman, Bernard McGuinness, and the county manager, Michael McLoone. Mr. Bruton told me that I was in a very unusual and privileged position to have the opportunity to design the square in my hometown. "Not many people can do that," he said. I was deeply aware of this responsibility. In a small town, you are not only judged by who you are but by your family—your grandparents, parents, and siblings. As I worked quietly on the scheme, I was conscious that it was not just the project but the social capital of my family that had affected how

I was designing it and how the public were receiving it. In some ways, that long lineage was allowing me the permission to be in this role.

At the same time, I was struck by who did not come to see the plans as much as by who did. Absent were those who had some historical grudge. There were people who I knew would never set foot in my house, not because of any grudge against me or my family *per se,* but their politics or perceived politics. In part to reach these people, I also regularly held formal meetings with the townspeople. I came to realize, however, that public meetings are not always the best measure of a community's input. The people who would not come to the exhibition came to the meetings, often with more critical feedback and with not having studied the plans. Generally, they wanted the plaza to be made smaller to maximize parking spaces. This issue became especially contentious, and the suggestions did not often make sense to me or adhere to good parking guidelines.

I came to realize that the points that people were making were rooted in the Irish Civil War of the 1920s rather than in the project itself. My grandfather had been on the side of the treaty with the United Kingdom, and many of the people who argued against the project had been on the other side. I heard one of those people, then in his mid-sixties, speak with great pride about the fact that he had never set foot in a particular bank in town, which was loosely associated with the other political persuasion than his. Rooted in the civil war were issues over land and landownership, and power struggles over who controlled and owned the land. I didn't take anything personally. But I listened carefully and considered their feedback, and potential motives.

I later came to read scholars like the architect, writer, and professor emeritus of urbanism, Witold Rybczynski, who confirmed my instincts that public meetings do not always result in better design. In his book *Last Harvest,* Rybczynski, eloquently documents the process of building a neo-traditional housing development outside Philadelphia, Pennsylvania.[32] When constructed, the main boulevard was 150 feet wide, an excessive width, which encourages faster traffic speeds and one can assume there is no shade from a tree canopy.[33] The desire to make these roadways so wide, against the principles of neo-traditional developments, could be traced to public meetings, showing both the capriciousness of the meetings while also exposing the failing of the designers for giving into such suggestions in the first place. In avoiding their own headache in public consultations, the developers were creating a longer-term headache for the users of the site.[34]

In the Diamond's case, the process was different mainly because of the individual conversations that I, as the designer, had with the townspeople, and the public meetings were advisory but had no binding result. In the sense that ethnographic fieldwork is about participant observation, this form of one-on-one engagement was a vital aspect of the design process in the Diamond project. These conversations complemented, rather than replaced, a more formal consultation process and public meetings. The method of cultivating individual conversations, albeit time-consuming, allowed residents a safe space in which to discuss their concerns and contributions more directly and with less confusion and more time to discuss. When a contentious issue came up in public I would follow up with a private meeting. The one-on-one conversations also preempted potential criticisms. I took note of the conversations and reflected on them. I later came to see their ethnographic value. The American Anthropological Association describe ethnography as involving "the researcher's study of human behavior in the natural settings in which people live."[35] In that sense, the Diamond was my first foray into ethnography.

Peace and Reconciliation

As I mentioned earlier, while I was designing the Diamond there was a much bigger political process taking place, for peace and reconciliation in Ireland. Until now, I have not said much about my hometown's being twenty miles (thirty-two kilometers) north of the Northern Ireland border. It is an area that relates to Northern Ireland as much as it is to the rest of the Republic. Older residents tell me that after the foundation of the Irish Free State in 1921–22, Inishowen opted to return to the United Kingdom, but the results were not implemented. After the Northern Ireland ceasefire of 1994, the European Union made funds available for peace and reconciliation in Ireland, known as the PEACE 1 program. The program incorporated counties on both sides of the border, taking in twelve counties with a combined population of 2.1 million people. The PEACE 1 program was an initiative of the then European Commission and the members of the European parliament, approved in 1995 to address the immediate legacy of the Northern Ireland conflict and "to support peace and reconciliation and to promote economic and social progress" throughout the region.[36] From 1995 to 1999, over 15,000 projects were funded in the Irish Northwest and €667 million was invested on both sides of the border, and the Diamond was one of them.[37] The Diamond was funded because a portion of the funding was earmarked for urban renewal, and this was one of the projects that was ready to build. In terms of European politics, I always wondered if John Hume and his wife, Pat, played any role in the Diamond, and I will likely never know. John Hume was a member of the European parliament (MEP) and member of parliament (MP) for the Foyle constituency just across the border in Northern Ireland. In 1998, David Trimble and John won the Nobel Peace Prize "for their efforts to find a peaceful solution to the conflict in Northern Ireland."[38] Most Saturday evenings, John and Pat would park their car in the Diamond and walk to the nearby Corncrake restaurant for dinner. Given that the funding was from the European Union, they must have known about the project. And although we discussed the project later, I never knew if they had any direct role in its funding.

Moving toward Construction

There are generally nine stages to a landscape architectural project—although, as I've noted previously, these do not always follow a linear path. The nine stages are: (1) project inception and initiation; (2) concept design; (3) design development; (4) construction documentation; (5) tendering (procurement); (6) construction; (7) construction management/administration; (8) maintenance; and (9) postoccupancy evaluation.[39] Heretofore, we have focused on the first two: project inception and initiation, and concept design. While the drawings were developed for public exhibition, when the plans were approved and the project destined for construction, we entered the next two phases—detailed design and construction documentation—which involved tendering the project to professional firms. As the project was funded by the European Union, certain protocols had to be followed. One was that the Donegal County Council was obliged to engage a professionally accredited firm, which I was not. So, the council appointed a local firm of architects (given the shortage of landscape architects) to oversee the project and instructed them to follow my design.

The issue was what to do with me. At first, the architecture firm, located about eighteen miles (twenty-nine kilometers) from my home, invited me to work for them—an arrangement that lasted all of two weeks. I had gone from being project designer to an intern in an architects' office. The architects, understandably, wanted to question all aspects of my carefully calibrated design. While that was all fine and good, they lacked my conviction and passion. Aspects of the proposal that for me were essential were dispensable for the architects. We clashed most over the central paving pattern. In my mind, it could not be made smaller to save money, or woven out of red bricks imported to the village. They wanted to model the design on St. Mark's Square in Venice, which bore little resemblance to this village. But more than the design passion, the architects lacked the local insight that I brought to the project, even though they were from a town so close by.

Fortunately, the ever-creative Richard Flynn, the County Council's area engineer, proposed a brilliant solution in a progress meeting. He had sensed my angst, although I do not recall articulating it directly, and in the meeting proposed that instead of working for the architecture firm I come and work for the County Council. That way, I would represent the client and have design control, albeit under his supervision. I jumped at the chance and started with the council the following Monday. The only way I could be hired directly by the council was as a laborer; as a public body, there were procedures in place for the hiring of staff. So, I joined the direct labor team. My official role was assistant to the project manager, but I would also be recognized as the designer. From then on, my relationship with the architects was different, and in fact improved. I was based in the local area office and occasionally made trips to the Council Headquarters in Lifford, about forty miles (sixty-four kilometers) away. I took part in all project meetings, which were held regularly between the architects, county architect, John Deaton, quantity surveyor, and project manager.

Finalizing the Design—in Brazil

During the final stages of design development, I took time away from the daily ins and outs of the project to spend a summer in Roberto Burle Marx's studio in Rio de Janeiro, Brazil. Burle Marx (who Jellicoe regarded as the top landscape architect in the world) had died a couple of years before my visit, but his studio continued. I wrote to Haruyoshi Ono, Roberto's successor, requesting I work in their office for a period to facilitate my study on Burle Marx and he replied, by fax, inviting me to come to Rio. After I arrived, Haru told me, "Oh, we don't want you to work! We want you to experience Roberto!" I spent the following months visiting Roberto's built projects, with free access to examine project files in the archive. When I finally had the chance to visit the Safra Bank on the Avenida Paulista in São Paulo, which had inspired the initial design of the Diamond, I was not disappointed. The Safra Bank roof garden was best enjoyed from above, and given the level changes around the Diamond, the view from above was important there too. I noted the complete lack of straight lines in the scheme, and the peculiar juxtaposition of colors, textures, and volumes. The embodied knowledge from that visit inspired me.

Enthused by Burle Marx's work, I struggled to complete Diamond's design. The original cross was not symmetrical, so should I attempt to correct this asymmetry and balance the new scheme with symmetry? Burle Marx had powerful curves that

Figure 27. Aerial view of the Diamond. (Photograph by David Lyons)

Figure 28. Safra Bank, São Paulo. (Photograph by Leonardo Finotti)

had little formal historical precedent—but were inspired by the landscape. This project was reinterpreting an historical symbol which was asymmetrical, and the issue was whether to reproduce the asymmetry or to correct it. Haru helped me to understand that the power and vitality of Burle Marx's work came from the lack of symmetry or, one might say, the presence of asymmetry. Through a form of fieldwork that combined the personal experience of precedents with talking with designers, I completed the design of the Diamond at Burle Marx's studio in Rio de Janeiro.

With an asymmetrical approach decided on, the distinctive interlacing pattern was difficult to draw accurately. Back in Ireland, I went through many variations, from freehand sketches to photographing the cross with slide photography and projecting the image on a wall then tracing it onto large sheets of paper. Next, I photocopied photographs of the cross and enlarged the photocopies, but the resolution was too unclear. Eventually, I went to the site with Danielle Moore and Máirín Nic Diarmada, who were both interns with the County Council, and made a physical rubbing of the cross using sheets of paper and crayons. That pattern was then digitized into a CAD drawing by Danielle.

When it came to the Christmas tree, I decided to integrate it into the pattern. Jellicoe had suggested the easiest thing to do would be to build an obelisk in the Diamond. Obelisks have a long history and often serve no practical purpose but have had various meanings and associations over time.[40] Those obelisks in the north of Ireland, such as in the Diamond of Derry-Londonderry, often represent British imperialism and war conquests. I opted instead to design a removable central stone in the paving pattern to hold the Christmas Tree, which became a sort of temporal obelisk for one month of the year. It was encased in a metal surround, and beneath was a six-foot- or two-meter-deep pit framed by a concrete wall. When I called the Belfast City Council, who had a similar arrangement, the engineer I spoke with told me it was important the pit was not too deep. He explained that the center of gravity of a tree is just above the roots; this is the strongest and most pliable part, which bears the brunt of the tension when a tree sways in the wind. If this part was buried

Figure 29. Skating on the Diamond. (Photograph by David Lyons)

the tree was likely to snap in the wind. This was knowledge I learned through others' experience, in the field.

Starting Demolition

The council decided to build the new project with what was called the direct labor force, a team of workers charged primarily with road maintenance, repairing potholes, and now the construction of the Diamond. Once construction started there were other relationships to navigate. The construction became another sort of fieldwork site that I needed to understand and adapt myself to. Anthony Cavanagh headed the construction team, and Don McGonagle, the newly appointed area engineer for Donegal County Council, served as project manager for the Diamond construction. Don was assisted in his job by Alexandra Lynch from Buncrana, who managed the council's suboffice and the payroll for the labor force. The labor force was headed by Michael Dander. A big advantage of using the local labor force was that they were also from the community. This was a form of active fieldwork in which members of the community were building the project that was for the community, and I was engaging with them on a regular basis.

To build, one must erase. The construction of the central plaza itself began with the demolition of the public toilets. It didn't take long, on a Monday morning, to knock it down using a JCB, a mechanical digger. Few people lamented the passing of the toilets, and almost everyone seemed impressed with how the town had been immediately opened up, confirming what older residents had told me about the previous demolition in the space. After the rubble was cleared away, the sunken part of the plaza had to be excavated and the levels set. The old concrete footpaths around the perimeter were taken up and the overhead wires undergrounded. Spare ducts were put in for services that had yet to be invented—such as the Internet. New pavements were constructed by a local firm, Seamus Toland, using reconstituted granite pavers. Basically, all surfaces in the Diamond were new. This process took about three months.

The council's team took care of all the demolition and the preparatory work for the new structure. The stone that we ended up using was a quartzite from West Donegal, supplied from a company called McMonagle Stone. The day that I visited the quarry to select the materials was rainy, which brought out the colors of the stone and aided in my decision. The stone walls that framed the central grove were built by Doherty master stonemasons from Ardagh in Clonmany—a father and two sons, using stone-building techniques perfected over generations. It was only through working closely with the stonemasons on the site, through careful exercise in scale and proportion, that we were able to work out the heights of the walls. Meanwhile, when the walls were almost completed, the specialist stonework was laid by Stone Developments to great precision. The individual stones, based on the CAD drawing produced earlier in the process, were crafted with such precision that there were four-millimeter joints between stones of the same color and seven-millimeter joints between the stones of different color. The joints were pointed with cement to seal and strengthen them. Liam's work on churches had influenced me to design benches inside the Diamond, which were crafted by a local joiner, Eugene McGuinness, from Oregon pine. These hand-crafted benches were described by Joy McCormick, Liam's wife, as "church feeling," which was not untrue.

Whins and Wild Cherries

Early in the design process, I suggested covering the Diamond in a monoculture of whins. *Ulex europaeus,* known locally as whin, and elsewhere in Ireland as furze, is an evergreen shrub that grows on poorer soils. Found mostly on hillsides and shrubland, the ubiquitous olive-green prickly leaves are adorned with bright yellow flowers that have a distinctive, fresh smell. The books tell us they bloom from March to May, but if the weather is mild, they can bloom for much of the year. They're used for fences, fuel, fodder, bedding, manure, roofing, walls, fencing, foundations, harrowing, hurleys, walking sticks, cleaning chimneys, dyeing, insulation.[41] They were used for decorating maypoles: "Whin blossoms are the usual decoration of the pole."[42] In addition to being an indigenous shrub that would bring the landscape of the countryside into the town, they had other benefits: they stayed green throughout the year; they would not require good soil or regular maintenance; and they and their prickles make them relatively vandal proof. There was just one caveat. The council's engineers pointed out another distinctive feature of whins: they burn easily. Often

Figure 30. The Diamond under construction, 1996–1997.

Figure 31. Sitting in the Diamond, ca 2001. (Photographs by Anne Whiston Spirn)

during dry periods in the summer, the hillsides would be ablaze with burning whins and heather. For me, that was not a problem since the whins regenerate quickly from their charred stubs. But the council were worried about liability issues.

In the end, I was only able to plant a handful of whins in the Diamond. I chose other shrubs and flowers that reflected the colors of the landscape, rather than being native to the landscape. An evergreen shrub, *Mahonia x media "Charity,"* replaced the whins, providing yellow flowers and year-round volume. I chose indigenous wild cherry trees for the central grove, each tree selected at a nursery near Belfast for its

personality and individual form. The nursery owners were no doubt happy to see me taking the runt of the stock from the nursery, but I had been inspired by another of Burle Marx's projects, the Copacabana Beachfront in Rio de Janeiro. Most public landscape architecture projects lean toward using uniform trees, as closely aligned as possible, but Burle Marx celebrated the particularity of individual nonnative *Terminalia catappa* trees for that project. In the Diamond, we also planted Irish ivy, *Hedera hibernica,* around the base of the trees. Not only would the ivy vines provide groundcover but they would clothe the trees, providing additional greenery during the winter months while the trees were bare.

Figure 32. Whins (*Ulex europaeus*), near Kinnagoe Bay, Inishowen, Co. Donegal. (Photograph by Alamy)

The trees were symbolically planted by the Inishowen Environmental Group and featured in the local newspaper. The same group organized an annual sale of saplings to encourage tree planting within the community. Later, when I showed photographs of the completed project to the landscape architect Ian L. McHarg, he quipped, "There's not enough trees. There should be more trees!" But there was little tradition of street trees in Irish towns, and they would be difficult to sustain with the rock under the surface. Later, as it turned out, this idea of celebrating plants from the local landscape turned out to be a contentious issue for several residents, and not one I was prepared for.

Official Opening

On June 6, 1997, Mary Robinson, then president of Ireland, whose mother was from the town and who as a child would buy ice cream in my family shop, inaugurated the newly built Diamond. Tom McBrearty, the retired headmaster of the local boys' primary school, provided the Irish translation for the plaque to commemorate the president's visit: *Máire Mhic Róibín, Uachtaráin na hÉireann, a d'oscail an láthair seo.* (Mary Robinson, president of Ireland, opened this place.) *Láthair* is an archaic Gaelic word for a centre, with no direct English translation.

While the official invitation came from the chairman of Donegal County Council, I was the one who organized the dates with the president's office at *Áras an Uachtaráin* in Dublin. I had tried to arrange this six months in advance and recall being admonished by one of the senior administrators from the council's headquarters in Lifford for my foresight. "We work on a day-to-day basis here," I was cautioned. At the opening ceremony, Michael McLoone, the manager of Donegal County Council, made remarks about the tension between history and the future. He said the design represented on the one hand an Irish, Celtic, and Christian heritage and, on the other, progress, renewal, and change.

Designer's Dismay at Dirty Diamond

In the immediate aftermath of the construction, I was quite happy with the structure and its maintenance. From my home, I would see neighbors such as Maud Smyth and her daughters Beatrice and Iris sitting and swinging their legs from the seats and eating ice cream as they waited for the bus home. I stayed in touch with members of the council regarding any major issues that came up with the Diamond. It was featured in the media, including as a case study and front cover illustration of *Landscape Design,* the journal of the Landscape Institute in London.[43] The project received an award based on "quality of design, construction, and general maintenance."[44] By 2001, however, the Diamond was already showing signs of wear and tear. Although I had developed a maintenance manual for the space, and a countywide version, too, the manual was clearly not being followed. The council had no

Figure 33. At the official opening of the Diamond by Mary Robinson, president of Ireland, June 6, 1997. (Courtesy of the *Derry Journal*)

maintenance budget and had a policy of delegating maintenance to the local population. When Anne Whiston Spirn, a professor of landscape architecture and planning at MIT came to visit, I asked the council to wash the structure with a power hose before her arrival, which they did. Later, when my colleague Niall Kirkwood and students came to visit, there was no one to ask to wash the stone, as I no longer felt able to ask anyone in the council to do that.

After construction, responsibility for maintenance of the Diamond was assumed by an association of well-meaning individuals called the Tidy Towns Group. They were assembled with the aim of succeeding in the national Tidy Towns competition, which includes various criteria for their assessments, including community involvement and planning (60 points); built environment and streetscape (50 points); landscaping and open spaces (50 points); wildlife, habitats, and natural amenities (50 points); tidiness and litter control (90 points); sustainable waste and resource management (50 points); residential streets and housing areas (50 points); approach roads, streets and lanes (50 points).[45] While the criteria are wide-ranging, it is telling that general category of "tidiness and litter control" earns the most points.

Embedded in the competition, as its name suggests, is an aspiration toward tidiness. I began to realize an underlying tension with aesthetics: The rural Irish landscape, upon which my design was built, is not tidy; it is wild. And the Tidy Towns Group does not like unruly plants. The whin bushes, in their eyes, were too scraggly. So too was the Irish ivy, *Hedera hibernica,* that was planted as a groundcover and that had started climbing and covering the trees. I heard complaints from the Tidy Towns members that the ivy was smothering the trees and killing them. I volunteered advice but was quickly rebuffed by an individual who said he was working on the maintenance for "the good of the town" rather than to further his career. By that point, I was living in Cambridge, Massachusetts, and had embarked on my doctoral studies at Harvard Graduate School of Design. Perhaps because he still lived in the town, he felt he knew better. I sensed that his feelings were shared by others in the group.

My connection with the Tidy Towns Group was finally severed when I came home to visit for a few weeks. I was especially frustrated with the maintenance regime and gave an interview to the local newspaper, which appeared in the *Inishowen Independent* under the headline of "Designer's Dismay at Dirty Diamond."[46] In the short term, the Diamond was washed, but in due course, the ivy was cut back to "protect the trees," they said. Later, the trees were cut down. They said they were old and had been killed by the ivy. The trees were replaced with much smaller, tidier species, which destroyed the feeling of balance and enclosure that the central grove was intended to elicit. At the same time, the plantings were changed to roses and geraniums, which look best in the summer months when the Tidy Towns judging takes place but are bare for most of the winter months. Meanwhile, the paving of the cross, with slabs just forty millimeters or 1.5 inches thick, was extremely delicate. I had placed bollards around the space to protect the stones from vehicles. The bollards were removed when various community groups, including the Tidy Towns Group and the Christmas tree group, insisted on driving on top of the paving. The stones, as I had predicted, were damaged. Finally, I decided to take the backstage. "Let them whack away," was how my grandfather would put it. At some point, the designer must let go and let the project have its own life, no matter how painful that might be for the designer.

At Home

This returns us to the crucial question for landscape architects raised at the start of this chapter: Who gives designers the authority to design for someone else? At a certain moment in time, my home community coalesced to give me the authority to design the Diamond. The executive powers at the county and national level accepted this consensus. They were generally happy with the result. As was I. The project sits within what might be termed "localism," and the (problematic) idea that only people from a community can design for that community. Gabriel Arboleda, for instance, described two basic premises to localism that this project fulfills. First, "that local social design practice should be carried out by local designers, and second, that designers should carry out their work by involving community participation."[47] Local, as Arboleda rightly points out, is a problematic term, and difficult to define, which is why I prefer "designing at home." Eventually, some townspeople felt the Diamond was too untidy for their tastes, so they proceeded to tidy it up. I underestimated the depth of the underlying renunciation of that "local" landscape in favor of an imported aesthetic model that looked more "orderly." The saying "Far away hills look green" (or in Irish, *Is glas iad na cnoic i bhfad uain*) speaks to this sentiment: that which comes from the outside is said to look better than that which already surrounds us. In this case, I posit that the fear of untidiness is tied to a—perceived—mainland European concept of aesthetics. One thinks of window boxes filled with geraniums in Bavaria, for instance, like the Diamond now has. Some might see this embrace of a tidier aesthetic ideal as illustrative of a gradual alignment of Ireland with the European Union. As a former market square, the Diamond was a working space, a place of trade and commerce, and probably never very tidy. While I regret not convincing the public of the merits of the scheme as designed, and I really don't like what the Tidy Towns Group did to the Diamond, in the end I can't complain. The community made a choice about how they wanted the landscape to look and function—and design, after all, is for people.

Here, I am reminded of Ingold's concept of a "taskscape." Ingold argues for an understanding of landscape rooted in a "dwelling perspective." Ingold tells us that the act of perceiving a landscape is to "carry out an act of remembrance, and remembering is not so much a matter of calling up an internal image, stored in the mind, as of engaging perceptually with an environment that is itself pregnant with the past."[48] The Diamond is the embodiment of this perpetual remembrance and human involvement as well as the interaction between land, nature, culture, and space. The landscape of the Diamond is an intuitive project, primitive, even. In fact, this project preceded most of my formal design classes, not to mention landscape history classes. When I later took classes in landscape history and theory—in Dublin—they drew heavily from Irish geography and Celtic mythology, an Irish approach to landscape history. On reflection, this project sits within those historical narratives.

The Diamond reveals an approach to landscape that is deeply rooted in human engagement. The Danish urban designer Jan Gehl tells us, "First life, then spaces, then buildings, the other way around never works."[49] The Diamond was a project driven by human concerns that then became spatial. At the same time, it was an experiential process, driven by circumstance and instinct and not by rigid professional protocols. As an undergraduate student, who had not yet taken any professional practice classes, I didn't know any other way to design. And as a form of activism, this project required a deep field-based approach. Diana Wiesner, the Colombian landscape architect and activist, insists that fieldwork is an essential part of activism;

it is from fieldwork that we understand the complexity of a landscape, like a sponge absorbing the "soul of the place."[50] At the time, I was studying Burle Marx and Jellicoe, who both prioritized landscape fieldwork, while my understanding of the Irish landscape was inspired by the playwright Brian Friel, neighbor and friend, known for the precision of his insights to Irish social and cultural life. Friel derived his insights from the minutiae of observations on Irish rural life. I was also impressed with the design work of the Irish landscape architect Martin Hallinan, and especially his award-winning Garden of Ireland at Expo 1990 in Osaka.

Inhabitants of Ballybeg don't often consult with architects; they engage with nonprofessional draughts people, known locally as "people who draw plans." And landscape architects are few and far between. Residents often associate landscape architecture with the gardens of "big houses," the remnants of a colonial past. "Landscape is different in Donegal. We inhabit it, (we are) not afraid of it," said Tarla MacGabhann, an Irish-speaking Donegal-based architect who trained with Daniel Libeskind in Berlin and teaches at Queens University Belfast.[51] I was quizzing MacGabhann on the accuracy of the Irish Gaelic word *tírdhreach* as a translation for landscape, given how *tírdhreach, a made-up word,* is not really used in everyday speech. MacGabhann gently but firmly reminded me how concepts of landscape grew out of landscape painting but concerns in Donegal are different, he said. In poor areas of the world, such as Donegal, we didn't have the luxury of landscape painting. "Landscape was a place we occupied. It gave us food or shelter, or we lived in it."[52] As a result, it is termed *tuath,* or countryside in Irish everyday speech rather than the official dictionary definition, *tírdhreach.* MacGabhann points out that to be out in the landscape is phrased *amuigh faoin tuath,* which literally translates as "under the landscape," suggesting something more immersive than *tírdhreach.*[53]

Reflections

In this chapter, I have introduced my process of designing and building a small village square, or Diamond, in the north of Ireland. Through the perspectives of resident, designer, contractor, and client—working at home—the entire design process was deeply rooted in landscape fieldwork, although I didn't recognize it as such at the time. The fieldwork challenged social hierarchies by engaging with various publics from within and without, from residents to presidents. The fieldwork approach was a complement to more statutory forms of public engagement. It's hard to say when the fieldwork started, but it extended through the design and construction process, and I started to disengage when a community group took ownership of the scheme. The building of the Diamond exposed some nuances between the understanding of landscape in an Irish context, where it has historically been a place of work and inhabitation. Will a more generic, tidy, landscape form prevail in today's more globalized world?

Looking back, I would not design a form like that today that so brazenly places a symbol in the center of the town. I don't see it as a mistake, but it is a product of a certain moment. In fact, my professional and academic training leaves me embarrassed by its upkeep and afterlife. The form of landscape fieldwork I did was deeply time-consuming, and I could do it because it was my hometown, and I was a student. Nevertheless, the process of the Diamond project has stayed with me all my

Figure 34. The Diamond. (Photograph by David Lyons)

professional life. Seeing the project from the multiple perspectives of designer, local resident, client, and contractor all at once required me to constantly renegotiate my own role and identity. Being able to shift between these identities also affected how I dealt with the various publics—including the townspeople, the contractors, the architects, the landscape architects, and planners, the politicians, and the president of Ireland.

To actively engage with a wide range of interlocutors across society requires a sustained period of participant observation. The landscape fieldworker will understand the concerns and values of ordinary residents as well as those in power. At that point, I was still a native of my hometown, not yet "corrupted" by academia and professional limitations and assumptions about the "right" way of doing things. There was a certain innocence and naïveté to the project, but it was built with an intimate knowledge of place derived not just from my own lifetime but accumulated over generations. The challenge it set me was how to possibly design for another community with a similar depth of knowledge or intimacy. After the Diamond, I was retained by the County Council and assigned to design other projects in towns and villages across the county, which I did. But I was always acutely aware of what I did not know about the needs and ambitions of the various communities I was working with. It was this discomfort that led me to want to know more about the skills that ethnographers have in gaining access to knowledge *away from home.* How to work in others' home societies with deference and respect? Such skills, I would argue, are found in tracking the complex relationships between people and land, which can be best understood through the methods of landscape fieldwork, which decenters, and perhaps recenters, our understandings of landscape itself.

2

THROWING BEANS

Operational Fields and the Prototypical Landscape

Every game takes place within the boundaries of its own spatial domain.

—Guy Debord, *Memoires*

Knock-knock! "Hello, how are you?" I asked to the confused face in the doorway. "Sorry to trouble you, but my bean landed on your kitchen table. May I come in?" I was driving around the Netherlands with three other designers in a rented Volkswagen for two weeks. Our task was following four thousand beans. We had thrown the beans on a large map—250 kilometers long (approximately 161 miles) and subdivided into 12.5-kilometer (7.5-mile) squares—and then set off to visit the epicenters of the random points where they fell. Never did a householder turn us away. Maybe the socially liberal Dutch liked having strangers enter their homes, or perhaps residents were captivated by the novelty of a bean falling on top of their house. Whatever the reason, the beans brought us to places we never otherwise would have visited, obvious and not-so-obvious.

The name of the firm I was working with, CHORA, was inspired by an inscription in the Church of the Holy Saviour in Chora, Istanbul. The word "chora" in Greek, *χώρα,* appears above an inner doorway of this, a former Greek Orthodox church turned mosque, then museum, then mosque again.[1] Raoul Bunschoten has described the meaning of the concept as "a field for the living, a field that links two realms; a threshold space between local and larger, global, conditions."[2] For Bunschoten, this transitional, liminal space where CHORA's design office sits—between imagination and building, intelligible and sensible—is a peripheral condition, an area in between the profane and the sacred, or what Bunschoten describes as "vortices in a landscape of change."[3]

For two weeks that summer, I was part of a whirlwind tour around the province of Noord Brabant close to the Belgian and German border. Our entire site, though, stretched from Rotterdam harbor—then the largest in terms of trading volume in the world—into the heart of Germany's Ruhr district. We were visiting what we called "beansites." The office had temporarily relocated from London to Rotterdam, and we had set up a temporary field office in the Berlage Institute Postgraduate Laboratory of Architecture, which was based there at the time.

Everyone in the office at that point, four junior architects and me, a landscape architect, was freshly out of design school. I had just graduated with a master of landscape architecture degree and a certificate in urban design from the University of Pennsylvania, where I went after completing my studies at University College Dublin. James Corner, my academic advisor at Penn, recommended I work in the office. He told me of all the firms in the world that he knew of, the one I was best suited to was CHORA. Their reputation was kind of mysterious and elusive to me, given that they had no built work that I was aware of. Still, I had read many of their texts as required readings for Corner's studios at Penn. CHORA's distinctive drawing style, incorporating social and cultural referents, was an inspiration for me and everyone in the office, as I assume they were for Corner.

In this chapter I describe the novel design method using beansites developed by CHORA and Raoul Bunschoten. The fieldwork-rich method begins with throwing beans on a map and visiting the points on which the bean falls, an experiment first conducted by Bunschoten during a workshop at the Aarhus School of Architecture in Denmark in 1997.[4] The method, celebrating chance and serendipity, brings fieldworkers to sites they might never have even thought of visiting. The technique has been employed not just in Denmark and the Netherlands but all over the world, in the United Kingdom, China, Thailand, Bahrain, and Australia—themselves perhaps seemingly random sites.

This method is not only an outstanding example of the relevance of landscape fieldwork and its integration with design and planning processes, but it is also an essential example of a form of fieldwork that is rooted in projection. The chapter demonstrates how teaching and research can be integrated with one another. Through the CHORA design process that we went through in several countries,

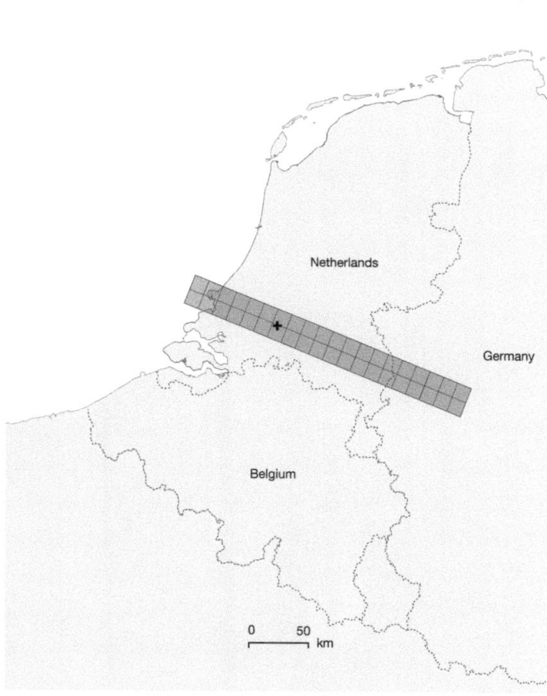

Figure 35. Sector E. (Maria Vollas)

I will explain a scaffolding that can be applied and adapted to other situations in space and time. In describing the design processes, I aim to demystify them. Before describing the steps, though, it's first necessary to explain some background terms and concepts.

> The world was born from Chaos, time emerged, infinite. Particles separate off from a primal mass, and form worlds on whose skin we live. Architecture helps us relate to this skin and thus to the larger imago mundi. It substitutes implicitly an evoked, imaginary space for a phenomenal experience. The fort becomes an instrument for the imagination through which this imaginary space is made into "sound." Strings with knives arc on the point to cut or play the particle/worlds encircling it. Architecture cuts—out of necessity—the skin of the earth, but simultaneously it makes the world into a "house of the soul."[5]

Figure 36. Soul's Cycle, 1992. Creating an incision creates a discontinuity on the skin of the Earth. (Photograph by Hélène Binet)

Figure 37. Soul's Cycle, 1992. The conditions the incision create are of more interest than the incision itself. (Photograph by Hélène Binet)

Terms and Concepts

CHORA's method assumes all the buildings we dwell in—as well as the programs and activities we engage and participate in—are sitting on top of the ground like layers of skin. Bunschoten describes the city as having two skins. The first skin of the earth is physical; it consists of disparate objects such as the buildings, bridges, cars, homes, information, motorways, oil gas pipelines, and even the weather. The second skin contains the hidden emotions and processes of a landscape.

Now imagine that the surface of a city floods, and the various parts lose their grip on the ground. The landscape's elements float on top of the flood waters. Then, the designer comes in and vigorously stirs up the elements with a giant wooden spoon. The stirring creates a vortex that liberates the ingredients of the built environment from their original location and function. After the stirring, some pieces remain as debris floating on top of the calm water, like flotsam, the remnants of a sunken ship. These pieces, now freed from their original situations and uses, can be recombined in radically new ways by the designer working with those who inhabit or have some form of agency over the site. This goal of recombination is what makes CHORA's fieldwork method so distinctive. It is a form of fieldwork that is intended from the outset to be projective, propositional, and collaborative.[6]

Fieldwork is key because it provides the content for a database of what are called "operational fields." These operational fields are the landscape processes that are taking place in any given place in time. Operational fields exist within the second skin of the earth, in the realm of the invisible.[7] Once, while searching for a precise definition, I asked Bunschoten if operational fields are like angels. I asked if they were, in essence, invisible forces floating in the air that not only influence the decisions people make that shape the land, but can also shape the earth directly. They surround us and influence us; one only must learn how to see them. This angelic metaphor is perhaps vaguer than others, but I find it helpful. Bunschoten agreed. Operational fields are never precisely defined by Bunschoten in his various writings. One of the clearest glimpses that we get is that they "contain the actors and agents that within their roles articulate the action-tendencies of proto-urban conditions and give them the potential to unfold."[8] Of course, "actors and agents" are problematic terms for they imply a hierarchy of agency, but everyone has an agency of some sort whether they be a multimillionaire developer or a homeless person sleeping rough on the streets. CHORA recognizes and flattens those hierarchies as part of the design process.

"Proto-urban conditions" are likened by Bunschoten to the tectonic forces that give form to the earth's surface, or as he puts it, "akin to the lava coming out of the thin crust of the tectonic shift running through Iceland."[9] These proto-urban conditions are not just products of the earth's processes, they can also refer to global shifts of capital, mass movements of refugees, or the devastating impacts of climate change that we see all over the planet today. In that scenario, proto-urban conditions refer to landscape processes, but operational fields can also contain the stakeholders that facilitate landscape processes.

Operational fields inform the prototypes that designers create. The *Oxford English Dictionary* defines "prototype" as "The first or primary type of a person or thing; an original on which something is modelled or from which it is derived; an exemplar, an archetype."[10] Prototypes can also refer to a "model," which the *OED* describes as "a full-size original of which a model is a representation on a reduced scale."[11] In the

CHORA sense, prototypes are formal but may include processes too. A CHORA prototype comprises a reassemblage of assemblages in a given point in time, putting together the flotsam, or debris, floating on the second skin of the earth. This process-based use of the word "prototype" appears to originate with CHORA, and deviates from the dictionary definition that considers a prototype more narrowly from the perspective of form.

CHORA's distinct methodology has four steps, which together are known as the "Urban Gallery," described as "a peripatetic instrument that supports the planning and design of complex environments in which many different parties and interests intertwine."[12] In the early version of the Urban Gallery that I worked on, these steps included database, prototypes, scenario games, and action plans. Each step is further divided into four substeps that I will explain below. The CHORA method was developed through the oscillation between teaching and the office. Bunschoten taught in the experimental and innovative environment of the Architectural Association School of Architecture in London's Bedford Square—what he described to me as a sort of sacred grove shielded from the reality of daily life. Another important reference was *Collage City* by Colin Rowe and Fred Koetter, whose concept of the bricolage is defined by the *Oxford English Dictionary* as an artwork, object, or concept developed by "appropriating a diverse miscellany of existing materials or sources,"[13] not all that different to the flotsam metaphor.

In the fall of 2000, Takuro Hoshino, an early and key collaborator with Bunschoten, explained to me over lunch at Mario's Café in Kentish Town, where the staff often had meals together, that they chose the number four for the four steps carefully. Four allows for many variations, but not an overwhelming number of them. CHORA's early work was influenced by cybernetic tools such as those developed by the cybernetician and inventor Gordon Pask. Pask had a teaching position at the Architectural Association at a time when computers were by no means the norm in architecture. Bunschoten recalls his first encounter with Pask, who was drinking a glass of white wine at the AA bar—he advised Bunschoten where to find an Indian restaurant. This encounter was the beginning of a relationship in which they later taught together. Pask was not an architect himself, but he collaborated with some of the eminent AA teachers. Pask also worked with Cedric Price, on his "Fun Palace for Joan Littlewood" project. Fun Palace, which was never built, was intended as

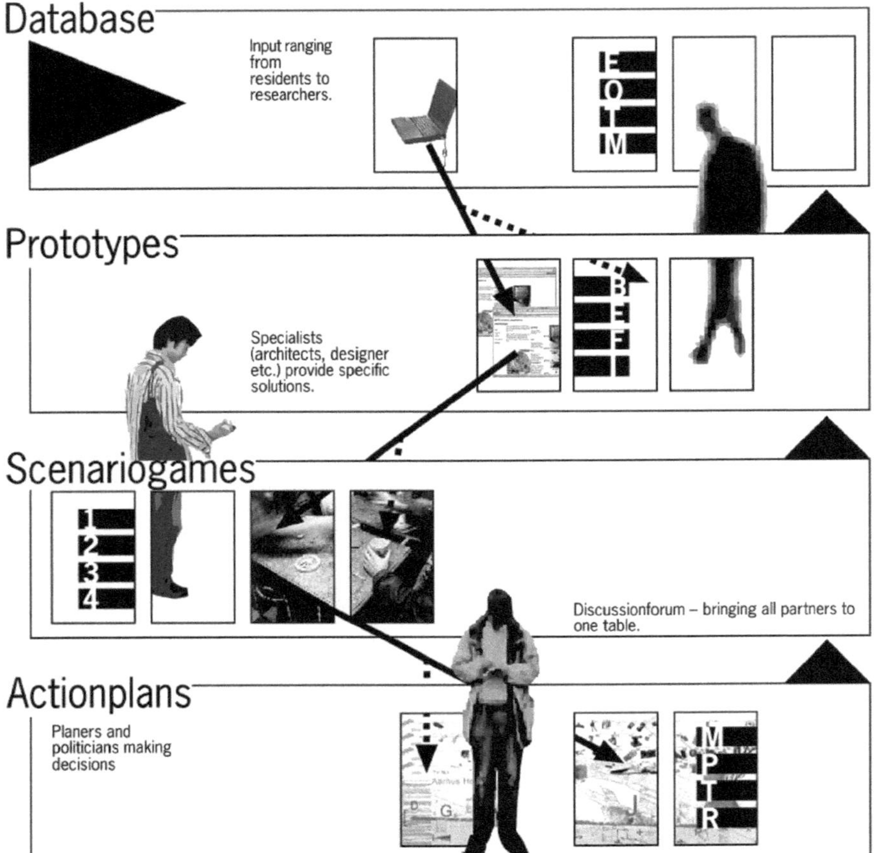

Figure 38. Four stages of the Urban Gallery. (CHORA)

a flexible space that its users could adapt and change. Designed as a "laboratory of fun" and "a university of the streets," Fun Palace was intended as participatory design project from the beginning. The lack of boundaries between scales, disciplines, and modes of describing landscapes, that was facilitated by cybernetics, helped the Urban Gallery become a digital tool. It anticipated the arrival of digital design, but the Urban Gallery became and has a different genealogy to, for example, GIS. By incorporating the second skin, the Urban Gallery aspired to integrate the soul of the city.[14]

Instructions for Fieldwork

Let's take a moment to look more closely at each of the four steps as well as what they mean for landscape fieldwork. These instructions are sketched out in *Urban Flotsam,* published in 2001 by 010 Publishers in Rotterdam,[15] and explained in later publications such as the Korean magazine *ANC Architecture & Culture.*[16]

Database

The process of building a database of operational fields is relatively straightforward. First, procure a large map. In general, the larger, the better (although, depending on the site, they probably would not be quite as big as the one-to-one maps that Lewis Carroll and Jorge Luis Borges might recommend).[17] Tabletop size is practical if the beans are thrown from above and not from the side. A room-sized map large enough to lay out on a floor and walk on top of is an optimal size since it makes it more feasible to drop the beans from a consistent height.

Second, find some beans. They can be black-eyed beans, broad beans, edamame, *haricots,* kidney beans, pinto beans—or even coffee beans or jellybeans. Beans are used because of their symbolic ability to grow into a new life. Each beansite is understood for its potentialities, rather than for what is there. For this reason, one should try to use beans with a generative potential over inanimate beans.

Third, decide on an indexing system. The map will need a subdivision, which will depend on the size of the map. A simple matrix usually works, following the existing subdivisions of the map, with a number-letter referencing system. It helps to use a consistent nomenclature with letters on the X-axis and numbers on the Y-axis. Thus, a bean may fall on, say, 4B, 5C, or 10D.

Fourth, throw the beans. The beans should show a relatively even distribution over the map. One may consider different development densities or types on the map, but not too much. Even if an area is low-density, it still has operational fields. Perhaps those landscape processes are more easily seen and deciphered because an area is low-density, or they get overlooked because they fall outside one's normal

Figure 39. Throwing the beans and marking the beansites. (CHORA)

disciplinary purview, timeframe, or comfort zone. Landscape processes, after all, are environmental, geological, and linked to the longer-term. If the beans miss essential sites, so be it, they cannot be moved. I always felt that one of the problems, if you follow the methodology too literally, is that you can miss the obvious. But the science of the method is serendipity. After the beans fall, and their locations are fixed, photograph the map with the beans to have a record in case any of them get inadvertently moved. Then mark the beansites on the map and determine and number the epicenters.

Fifth, once they are marked, visit the beansites. Note that visiting beansites involves a certain amount of discretion, if not creativity, on behalf of the fieldworker. For instance, the epicenter is rarely very precise. Perhaps the bean fell on an entire neighborhood. Perhaps it fell in a garden or in the middle of a roadway. Often, it is hard to determine whether it fell inside a particular house, or outside. Rarely can one say with any certainty that it fell in a specific place within that house, such as a kitchen table. Maybe the bean was suspended in the air or fell underground. Till now we're assuming the beans fall on the plan, but what if we consider them sectionally? The fieldworker will thus choose the exact point in space, and time, to visit. As a result, the precise spatial coordinates are usually determined from the fieldwork experience itself and informed by field conditions, not to mention the biases and knowledge of the fieldworkers. One house may seem more inviting than another from the outside, and that might lead the fieldworker to choose it, or not. Or perhaps the fieldworkers are aware of an overall lack of beans dealing with food and sociability, which might lead then to ask someone about their kitchen table. What might seem random is filled with opportunities for personal choice—and prejudice. Still, it provides an overarching framework for bringing fieldworkers to attend to the less-obvious dimensions of the landscape.

Sixth, once you are at the beansite, determine a miniscenario. A miniscenario is a set of standard questions posed at one point in time, which follows a narrative constructed *in situ* about the range of actors and agents, topics, and scales implicated in the site under study.[18] The miniscenario is, Bunschoten tells us, "written on the page of everyday life."[19] Construct this miniscenario for the site around four basic processes: erasure, origination, transformation, and migration (EOTM). These processes are taken from the stages of growth of a plant from seed. First, there is the soil;

the seed is sown; the seed grows; and seed blows away. The metaphor is based on the idea of the city as a garden. Designers and planners are like gardeners. Throwing beans on the map is like sowing seeds in the garden. One thousand beans could be one thousand seeds. Bunschoten defines these four processes in this way:

> *Erasure:* What is removed or taken away from a site?
> *Origination:* How did this site emerge?
> *Transformation:* What is changing in this site over time?
> *Migration:* What is moving through the site?

Figure 40. A miniscenario. (CHORA)

The object of the miniscenario is to identify operational fields based on the proto-urban conditions and the actors and agents that are affecting the site at a particular moment in time.[20] In principle, there is no causal relationship between the relationships, although they can flow from one to the other in the narrative. They provide a sorting mechanism to compare different beansites.

Let's imagine an example, based on the bean falling on the kitchen table in the opening vignette in this chapter.[21] Say, after answering the door, the householder leads you to the kitchen and the family is sitting around the table having their evening meal. This intimate encounter involves a family of four and a grandparent, and a broader set of proto-urban conditions that facilitate the interaction and give it structure. The miniscenario unearths the landscape processes flowing through that table in time and space. You can break these down by the EOTM method. The following miniscenario is guided by an example in the *Urban Flotsam* book.[22]

ERASURE

First, what has been *erased* during the meal? Clearly, each mouthful of food that the five family members take, erases food from the table. The potatoes, vegetables, and fresh fish are almost finished, but they invite me to join them for dessert of freshly baked Dutch apple pie and coffee. Even though we have so many other miniscenarios to do that evening, I accept their kind offer and use it as an opportunity to tell them what I'm doing and know more about them.

ORIGINATION

I explain to the family that from the erasure at the table, I ask what has begun that is new? What has *originated* from the family meal? I can see more than one original condition that has emerged as a direct result of the erasure of the food. Most obviously, what was once a table laden with food has now become a dirty table. As family members finish their plates, they bring it to the sink for washing. That day, it is the teenage son's turn to do the dishes. He's rushing to go to a football match. While the family continues to ask me more about what I'm doing, Jeroen washes the dirty plates and cutlery in the sink. The family are energized from the meal.

TRANSFORMATION

As Jeroen washes, the dirty plates *transform* into clean plates. Jeroen sets the clean plates and cutlery aside to dry on the draining board. I remember thinking it won't take long for them to dry in the warmth of the summer evening.

MIGRATION

The dirty water flows down the drain, presumably *migrating* into the wastewater disposal system of the region. Meanwhile, Jeroen puts the food scraps into the compost bin—they tell me it's collected weekly by the city.

OPERATIONAL FIELDS

A generic everyday interaction of a family, just described, that will be familiar to many, reveals several proto-urban conditions. First, it allows you to determine who are the actors and agents. The agents are the people who have agency over the dinner table, and the actors are those who don't have much control over it. Of course, these are debatable categories and problematic for a range of reasons as they reinforce hegemonies. But let's take them at face value for now. In this case, the agents are the cooks and, to an extent, the gardener who tended the organic apples for the pie. The cook is also implicated; they told me that that day the grandmother cooked the pie using a family recipe. The actors, of course, involve the family too, not to mention Jeroen who washes the dishes. The person who washed the dishes is an actor, too.

So now, you have made a preliminary list of who is involved. Next, address what processes—or operational fields—are in play. "Operational fields," Bunschoten tells us, "are further and more precise analyses of chosen phenomena based on mini-scenarios. They describe the underlying mechanism of the observed processes and the related actors and agents as driving forces to those processes."[23] In this case, the processes might include the backyard gardening and the growing of chickens. The climate that was so perfect for apples. The freshly baked apple pie was baked in-house, and the ingredients included eggs from the family's chickens, apples from the garden, and salted butter imported from Ireland—because of the common market and free movement of goods within the European Union. The coffee, meanwhile, was a unique blend from Sumatra, a former Dutch colony.; the selling of coffee; the socializing between friends; the brewing of espresso and frothing of milk; the farming of cows for milk production. Look closer, and you see that the teak used to make the table probably came from Southeast Asia, Indonesia perhaps, a former Dutch colony. A host of local and global processes—operational fields—have enabled the intimate family dinner.

It is all too easy to overthink the miniscenario that one writes on the site. To limit any confusion that might arise, set a time limit. Four minutes is usually good. Five minutes should be the maximum allocated. Then move on to the next beansite. Repeat this procedure until you have created a database.

I left the house after dessert, feeling a little guilty for taking longer than the requisite time for the miniscenario. When I later met with my coworkers, they had amassed many more miniscenarios during our evening of fieldwork. Back in the field office, we recorded our miniscenarios on cards and identified operational fields. In all, we amassed several hundred miniscenarios over our time in 's-Hertogenbosch, and each miniscenario yielded on average three to four operational fields. These thousands of operational fields were recorded in a database, and that database was available for the next stage of the Urban Gallery—the development of prototypes.

Prototype

Once you collect a critical mass of operational fields into a database, you can begin to recombine them in new and different ways to make prototypes. Prototypes are process-based precipitators of further processes or events.[24] Note that four basic criteria inform prototypes: branding, earth, flow, and incorporation (BEFI). Imagine them as "steppingstones" on top of the second skin of the earth.[25] Bunschoten describes these four processes in the following way:[26]

Branding—The programs, or activities, which contain conditions of cultural production, identity creation, images, and narration, naming, marketing, memory, remembrance, and value-creation.

Earth—Elements relating to the physical aspects of the prototype: the land, water, air, natural processes, ecological issues, biodiversity, but also land/territory and spatial organization.

Flow—The flows through the site, including above- and below-ground; the goods, information, money, traffic, people, and waste.

Incorporation—Programs altering and creating the institutional, legal, managerial, political, and social frameworks for a project.

Create prototypes by combining at least two operational fields; indeed, the networking of operations is what enables the intervention to be prototypical. Ciro Najle, dean of the School of Architecture and Urban Studies at the Universidad Torcuato Di Tella in Buenos Aires and former director of the MA in Landscape Urbanism Program at the Architectural Association, describes a prototype as "a contingent assemblage of relationships in a complex evolving system."[27] The development and construction of a prototype should respond to changing conditions rather than following one absolute, fixed plan (although they may need to be guided by a plan, particularly in the case of professional or legislative processes). The operative agency of the prototype is crucial, but form and appearance are important too. As James Corner writes, arguments for the staging of open-ended (and thus prototypical) interventions "do not make sense in a world without specific material form and precise design organizations."[28] Such prototypes may be temporary or long-term, but those made up of operational fields that are already existing in a site are more likely to succeed in that given site. We must recognize, however, that there is no obligation on the operational fields to lead to a prototype. They arise from different epistemologies. The database is knowledge that may, or may not, inform a prototype. Design is not a linear process.

The above definition, of sorts, taken from the book/manifesto *Urban Flotsam,* indicates two main criteria for a prototype. In essence, it must be open to adaptation or change and, by generating conflicts and further reactions, it must be open to proliferating.[29] A prototype is thus a design intervention that causes transformation, and which is itself open to adaptation and multiplication. Prototypes, as "the first or primary type" of something, cause effects like ripples disseminating from the point of contact when a stone hits the water. The ripples facilitate ongoing catalysis as new possibilities and opportunities unfold and reverberate over time and space.

As a design strategy, there are historical examples of what could be called prototypical. After all, most design interventions are inherently prototypical because they are intrinsically relational and cause effects in the surrounding landscape. One striking case that Bunschoten tells me about is All Souls' Church at Langham Place at the head of Regent Street in London. Unable to purchase the land to complete his preferred layout of Regent Street, the architect John Nash built the church in 1822 as a future focal point and terminus for the street. At that time, the view of the church was blocked by an estate whose owner refused to sell. By setting up the conditions to initiate change, over the next twenty years, the street was completed as Nash wanted, but without his direct intervention.[30] The land was purchased, and the two ends of the street met, like digging a tunnel through a mountain. By offering the vision, and the tangible image of a preferred future, the church acted as a catalyst for change.[31] In that sense, All Souls' Church can be called a prototype. Nash had set up the precise conditions for future actions.

Another example of a prototypical landscape is the Al-Azhar Park in Cairo. The Al-Azhar Park, by the Aga Khan Development Network (AKDN), was designed to act as a catalyst by initiating a range of community-based projects aimed at improving living conditions in the surrounding Darb Al-Ahmar district.[32] By providing ancillary processes such as microcredit and employment programs, as well as facilitating building renovations and historic preservation, the park is an exciting example from the perspective of prototypes.[33] Over one million plants and trees were propagated in the on-site nursery for use in the park, and later for sale with the hope of greening the notoriously ungreen Cairo. Although the implementation of the project is not as successful as it might have been, and the neo-Islamic design somewhat disappointing due to its attempt to recreate the past rather than design a new future, the process of the park's design in many respects compensates for the frustrating formal result. One can think of it as a prototypical landscape for the range and quantity of actions it precipitated through community engagement and so-called development.

This fundamental point about prototypes resonates with what the architect and historian Kenneth Frampton called for with the notion of a "catalytic architecture" that would influence the surrounding landscape. Frampton writes: "As architects, we need to conceive of future urban interventions in such a way as they have a wide-ranging catalytic effect for a given amount of investment."[34] This implies that one can effect a broader landscape change by making precise design interventions. In this sense, the desired intervention is not in the form of the architecture itself, but in the wider reactions it brings about. Charles Waldheim writes of landscape urbanism as "landscape conceived and designed as the primary ordering element for decentralized urbanism."[35] In this sense, the prototypical urban landscape may reclaim the agency of landscape architecture and urban design from planning and architecture as the "form giver" of society. This understanding of the role of prototypes leads to more synthetically driven, real-world, flexible form of scenario-based design than a more fixed or determinate form of landscape architecture or planning.

Take, for example, the World Trade Center in Dubai, United Arab Emirates, designed by John R. Harris and Partners and built in 1979. The building acted as a catalyst for the further transformation of the desert landscape to the city we now know as Dubai.[36] In this sense, the Dubai World Trade Center has been the precipitator of an urban landscape. Likewise, in the very different setting of the favelas of Rio de Janeiro, a project initiated by the city of Rio and funded by the Inter-American Development Bank has had a transformative impact on the urban landscape of some favelas or low-income settlements. These favelas, often situated between the country and the city, are "soft grounds," communities undersigned by professionals in contrast with the hard straight lines and asphalt of the formal city. In the favelas, public spaces have been introduced as remedial devices, with the intention of making them more inhabitable. An Argentinian architect, Jorge Mario Jáuregui, spearheaded the Favela-Bairro project, which resulted in many new prototypical spaces.[37]

These ideas resonate with other landscape architecture thinkers and practitioners who see spatial interventions as a way to bring about broader social, environmental, and aesthetic changes. For example, James Corner's ambition for an understanding of the urban landscape as "a thick, living mat of accumulated patches and layered systems, with no singular authority or control."[38] The architect Stan Allen credits much of his interest in landscape processes to the influence of Gregory Bateson, an eclectic writer who argued that an "ecological understanding must be ecological" in reference to the use of metaphors to describe the city.[39] In this case, Allen is referring to the word "ecological," and as George Lakoff and Mark Johnson tell us in the title of their book, there are many "metaphors we live by."[40] They reveal an idea of landscape that expands upon a rhizomatic condition.[41] Add to this the cultural geographer David Harvey, who argues that urbanization consists of a "distinctive mix of spatialized permanences in relation to one another" and that efforts should be directed into intervening in these processes and their relationships as a way to bring about change, rather than into form *per se.*[42] The Yale School of Architecture professor Keller Easterling tells us in the introduction to her book *Organization Space,* a prototypical approach to design is an appropriate response to the network society: "to truly exploit some of the intelligence related to network thinking, an alternative position might operate from the premise that the real power of many urban organizations lies within relationships among multiple distributed sites that are both collectively and individually adjustable. . . . It pursues a fascination with simple components that gain complexity by their relative position to each other. For example, it is possible to understand sites as separate agents that remotely affect each other—that is, the way one can affect point C by affecting points A and B."[43] Easterling situates prototypes within network theory and the rhizomatic society, as discussed by Manuel Castells and Saskia Sassen, among others. The use of prototypes in design is something that is responding to the needs of the present age, of risk-based society, individualism, globalism, and globalization.

"Landscape urbanism," says James Corner, "implants a new potential in a given field through the orchestration of infrastructural catalysts—infrastructures that perform and produce, or 'exfoliate' effects."[44] James Corner Field Operation's work has been characterized by phasing, such as parks that are implemented over twenty or thirty years, the design of the park responding to changing dynamics over time; for instance, JCFO's plans for Fresh Kills in New York City. However, Corner recognizes that there are laws, legislative processes, and protocols to be followed and that fragile politics and egos may not allow a project to get beyond its first phase. He says that the first phase ought to be precise enough to be able to generate the next few stages, even if the subsequent stages are never officially commissioned. In essence, although he is not using the terminology, Corner is calling for a prototypical landscape—and a proto-political one.[45] Prototypes offer a strategic, canny, and fluid approach to spatial design and planning—and perhaps are best used in conjunction with more conventional design and planning methods, rather than in opposition to them. The takeaway is that landscape architects need to work with such effects of adaptability and proliferation if they are to have any effect over them.

When a prototype consists of operational fields that are already existing in a site, we can deduce that the prototype is more likely to succeed in that given context. Having said all this, however, let's also acknowledge that design is not a linear process, and that arriving at prototypes may bear no obvious relationships to the operational fields. And that is okay too. "A prototype may also be a spaceship and just land," said Bunschoten, "like waking up from a dream."[46]

Scenario Games

Having developed a set of prototypes, the next step is to develop a scenario game to simulate and test the conditions through which each prototype can best be initiated and developed. Scenario games involve the same four steps as the miniscenario: erasure, origination, transformation, and migration (EOTM).

First, determine the players. Bring the various participants together around one table, along with a moderator, known as an animator. The animator is responsible for keeping the game running.

Actual stakeholders who would be impacted by a potential design process are invited to participate in person as much as possible. Playing games around a game is a great way to facilitate conversations between, say, residents and a mayor or maybe even a developer—individuals with different aims and perspectives. When conducted in an academic setting, a scenario game also has the capacity to work as a leveler between students and teachers. Here, I am reminded of Günther Vogt's strategies for putting himself on the same level as his students.[47]

It's a form of flat fieldwork, with hierarchies flattened. You may find that the atmosphere in the room shifts when the game starts. The players may loosen their shirt buttons or roll up their sleeves. In the process, perceived hierarches start to break down. Conversations and lines of communication that were not possible before, become possible. Even when all the stakeholders cannot participate in person, the points of view of the stakeholders can be simulated through actors adopting their role in the game (four players is ideal).

The game is both a platform for testing ideas, as much to link and mediate between disparate groups.[48] Participants bring in the prototypes they have already developed to test them through miniscenarios, as described earlier. These scenarios don't need to be rooted in fact. In fact, they may be fantastical. The objective is to cultivate

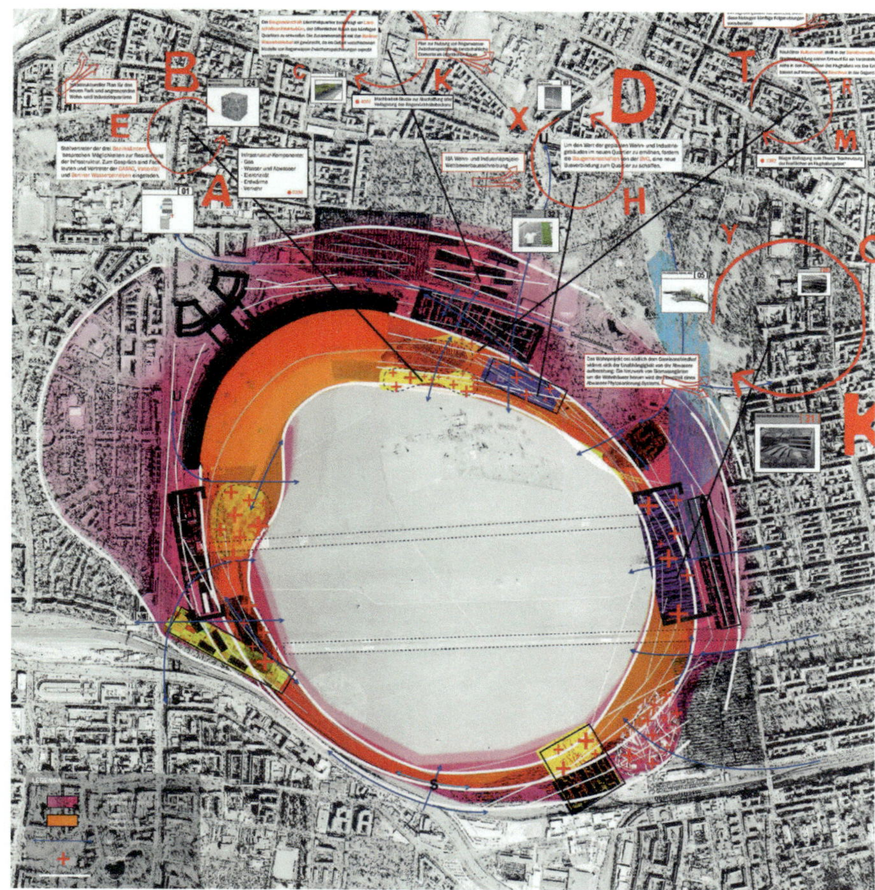

Figure 41. Tempelhof, Berlin. Denotes an energy positive district. (CHORA)

and give form to the imagination. Each player takes turns with the EOTM method. First, what has been erased? Then, they turn to origination: From that erasure, what new has already started? From that new beginning, what has transformed? Finally, from that transformation, what has migrated? Each player may have a different version of events. The point of the game is to reveal and articulate these multiple perspectives. As with the miniscenario, the game is best when limited by time. The role of the animator is to keep the game moving and to keep it focused on the testing of the prototype. So, if the game goes too far off-topic, the animator brings it back.

Imagine the following sample scenario. In addition to the animator, the game leader, around the table are

1. Local resident
2. Schoolteacher
3. Mayor
4. Newspaper reporter

You present the players with several prototypes for designed, multipurpose spaces in the city of 's-Hertogenbosch, the capital of the province of Noord Brabant. The first is a "cof-fire garden," a mix between a fire station, a coffee shop, and a garden. Another prototype is for a school farm, an urban farm managed by school children. The third is a church center, a mix between a shopping center, church, and school. Let's begin with the cof-fire garden. The idea of combining the fire station with coffee shop and garden came from the observation that, in between fighting fires, firemen and women could run the coffee shop. It provides them with something to do in between emergencies. The cof-fire garden contains the following operational fields, which should be integrated into the game: the extinguishing of fires, meeting friends over coffee, and the garden as a space for living.

Round 1

1. Resident: *The firemen and women have been temporarily "erased" from the cof-fire garden while they go to rescue a burning building.*
2. Schoolteacher: *The burning building is the school, which is now a ruin. What are we going to do?*
3. Mayor: *Don't worry. I will turn this ruined school into a better school building than it ever was before. From this fire a new transformed school will rise from the ashes.*
4. Newspaper reporter: *My story on the school fire is selling like hotcakes from local shops. It seems like it might not have been an accident after all.*

Round 2

1. Resident: *The shelves in the newsagents' shop are empty because so many people are buying the newspaper with the story on the school fire.*
2. Schoolteacher: *As a result of the empty shelves, I write my own inside story on the fire and set up a bank account for donations.*
3. Mayor: *I find another property for the school to move to while the schoolhouse is being renovated. It's a former bank.*
   * Animator intervenes because the story has moved away from its original intention of testing the feasibility of a cof-fire garden.
4. Newspaper reporter: *The firemen and women have migrated away from the cof-fire garden, and there's no-one to serve the customers.*

Round 3

1. Local resident: *The profits have been erased from the cof-fire garden.*
2. Schoolteacher: *The coffee shop closes, and the cof-fire garden becomes a fire garden.*
3. Mayor: *This fire garden becomes a public park, a place for people to learn about fire, and how to manage the fire. We'll have fire and water in the garden.*
4. Newspaper reporter: *This will be a very compelling story in social media.*

And so on. Usually, games are cut after a certain time limit, often after six or seven rounds—ten rounds being optimal. The idea is to test the feasibility of prototypes as well as their ability to adapt and proliferate. The game revealed the rather obvious flaw in this idea: Who runs the coffee shop when there's a fire? The game offers an alternative prototype: a fire garden, which would be housed adjacent to the fire station. The fire garden is imagined as a public park and an educational space in which the public is informed about fire and how to control it. With more time, the game will test the feasibility of the fire garden, perhaps combining it with other operational fields from the database and refining the idea further.

The Urban Gallery shares some common intentionality with Paulo Freire's *Pedagogy of the Oppressed.*[49] Freire rejects the so-called "banking" notion of teaching, this is where an all-knowledgeable professor reinforces his or her power by treating students as "empty vessels" that need to be filled. Such an approach to teaching sees the teacher/professor pouring knowledge into these empty vessels; and the act of domination hinders the capacity of the oppressed to learn. Instead, Freire proposed a model in which students don't just get filled with knowledge which they then repeat, but they create their own knowledge informed by the world they live in. In this scenario, students and teachers are coproducers of knowledge.

The Brazilian "theater practitioner" and political activist Augusto Boal, inspired by his compatriot Freire, took a similar perspective with *Theater of the Oppressed.*[50] Boal invites participants to adopt a role in a public role-play. Spectators become "spect-actors." In becoming spect-actors, individuals are liberated from their own perspective, biases, and hesitancies and allows a more direct and low-risk form of exchange. One can really say what one thinks, when representing a mayor, or developer, for instance, precisely because it is not the individual speaking, it is the character. But more than this, it helps everyone to recognize the range of complexities involved in decision-making. While serving as a city councilor in Rio de Janeiro, Boal developed the allied concept of legislative theater, which was directly focused on developing laws and policies to solve problems. In legislative theater, "people who are suited and booted have to take their suit away; they have to get rid of their title, they have to listen to the real raw truth of what's happening";[51] a "flow of power" is opened between citizens and the institutions of government. The method resulted in several new laws in Rio de Janeiro during Boal's term of office. Many of these same flattened hierarchies inform and empower the CHORA scenario games. Yet, it's important to recognize that the coproduction of knowledge is not the rejection of professional knowledge or expertise. It is the recognition that professional knowledge sits alongside embodied knowledge, and a negotiation must take place to translate the professional knowledge to a particular landscape, and vice versa. Like the scenario games, the Theater of the Oppressed is managed by a neutral facilitator, called the "joker." The same role is called a "animator" in the CHORA scenario games.

Pre-Texts is a similar creative system, inspired by literary practices in Latin America, and developed by Doris Sommer, director of the Cultural Agents Initiative and a professor of Romance languages and literatures and of African and African American studies at Harvard University. The method is used to encourage the close reading of challenging texts through forms of visual or other art making. The Pre-Texts website outlines a five-step process that takes place after an initial ice-breaking session. A volunteer reads the text aloud while participants create book covers for the text. Next, participants ask the text a question while they publish it on a clothesline—inspired in no small part by the Latin American practice of *literatura de cordel.*[52] Then, participants make art from the text. They form a circle and ask what did they all do? And, lastly, they "go off on a tangent" exploring extreme possibilities.[53] The main purpose, as with the CHORA scenario games, is to open space for imagination; this is done by decentering the process from individuals so they become part of a larger creative process.[54]

Action Plans

Using the data from the previous three steps, prepare a "dynamic masterplan" for the implementation of prototypes as well as the potentials for proliferation and adaptation of the prototype. As mentioned earlier, Bunschoten describes these as two essentials for the prototype.[55] The plan speculates on the potential interrelationships between prototypes and their application in space and time. An action plan also identifies the likely stakeholders as well as the operational fields that are embedded within the prototypes.[56] A dynamic masterplan is a plan whose coordinates are not fixed—a plan based on anticipation. The idea of anticipation is open-ended, a bit like a goalkeeper aiming for the ball: They know it's coming but they're not exactly sure where it's going to land.[57] To this end, the plan is a critique of normative master planning, which tends to have a fixed, definite result (although admittedly, this can be an overly simplistic reading of planning too).

Urban Curation

As noted earlier, the combination of the four-step process described above come together as the Urban Gallery. The name suggests an art gallery, which makes the designer like a collector looking around at the various works of art. Tellingly, the Urban Gallery is managed by a process referred to as "urban curation." The urban curator is different from an urban planner in the sense that to curate implies an active engagement with the design process. Curation usually involves the reassembly

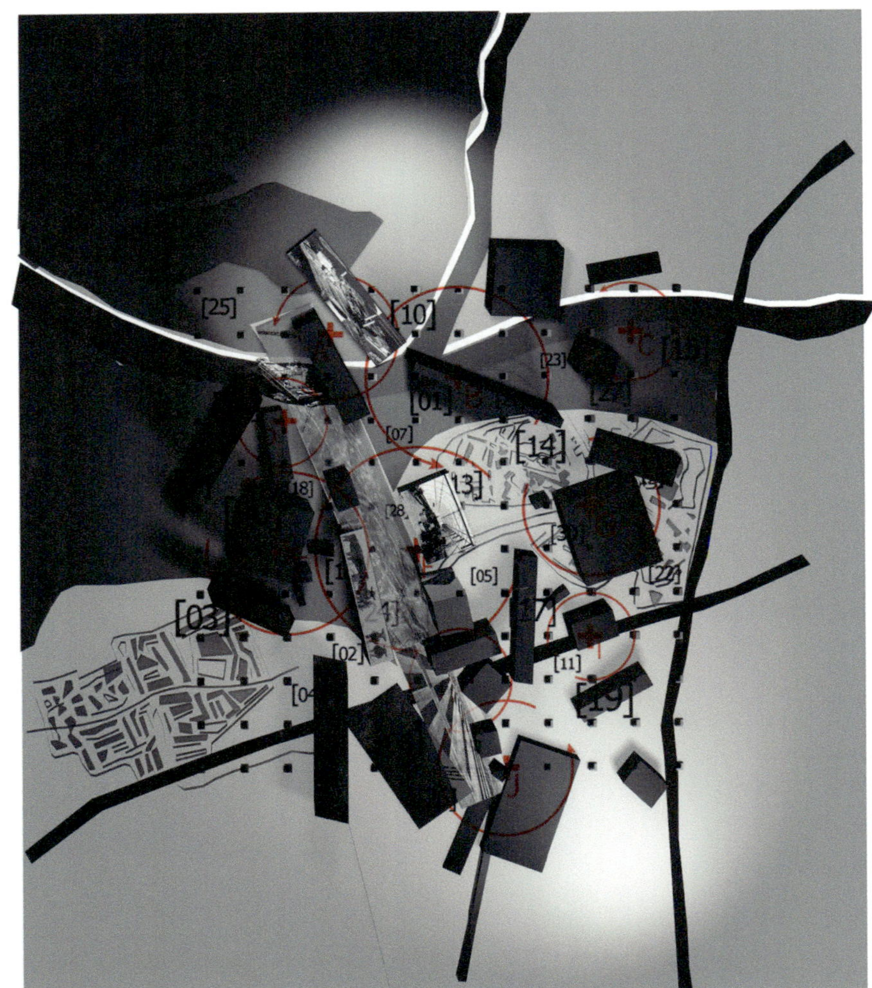

Figure 42. Sector E, Project W. Dynamic masterplan consisting of protoypes. One of them was using waste heat from a coal-powered power plant for heating greenhouses. (CHORA)

of an assemblage of different parts to create a new reality. In this sense, curation is an active and simultaneous engagement with design. The curator is a designer working with the latent connections between environmental, social, and aesthetic elements or ecologies, and reconstructing them. It is about taking care of humans and the human spirit, and objects and processes. Niall Kirkwood suggests that "stewardship" is perhaps a better word than curation and points out that landscape architects have long claimed such a role in cities. Kirkwood describes landscape architecture as a discipline that has "from its inception, engaged in a critical way with society, ameliorating the environment and proposing courses of action that reconcile the interests of profit with the public good."[58]

Games as a Form of Participation in Landscape Architecture

If people are part of landscape, then understanding human interactions is an essential part of landscape fieldwork. Giving a voice to the user is not about giving up the agency of design: On the contrary, landscape architecture can be more impactful if the needs and desires of the user are understood and considered as part of the design process. The question is how to interpret these needs and how to work with them in a constructive and imaginative way? By generating interactions between parties that otherwise might not meet and helping to break down perceived hierarchies between the various stakeholders, games can serve as a useful method in the designing and planning of landscapes. Games, in this context, are less about entertainment and more about opening possibilities for conversation, collaboration, and design. As the architect and game designer Ekim Tan writes that generative city-gaming has the potential to be a laboratory that exposes professionals to stakeholders in a process of co-creation; "Plans may therefore develop continuously while they are co-created and shared with the public through open and interactive communication."[59]

One limitation of games, in a professional context, can be their association with playfulness; it is difficult for a mayor or other public official, for instance, to be seen spending public funds on apparently frivolous activities such as games. Planning and design are supposed to be serious and sober activities, after all. For this one reason—the avoidance of the public display of human emotion and play—the scenario games rarely get to the next level. An article in the Danish daily newspaper *Dagbladet Information,* in October 2002, describes CHORA's throwing of four thousand beans on a map of Copenhagen under the heading of "*Da mr. Bean kom til byen,*" which translates as "Mr. Bean comes to town," inviting a comparison with the British sitcom starring Rowan Atkinson. The tone of the article is a case in point. While factually correct, and there is undoubtedly nothing libelous in the text, the article reads as dismissive of the concept of beansites and scenario games. The article begins:

> *I byen*
> *Hvide bønner eller brune?*
> *»Hvide.«*
> *Blev de virkelig smidt ud?*
> *»Ja!«*
>
> In the city
> White beans or brown?
> "White."
> Did they really get thrown out?
> "Yes."

But games need to be taken seriously as planning and design tools. They're valuable because they bring the intentionality of planning and design to the public and make the ideas exciting and palatable, and communicable. Most importantly, games offer a serious way to understand the present and to imagine the future. If we don't or can't imagine the future, especially preferred futures, we have no hand in shaping those futures.

Reflections

Beginning with the example of a bean landing on a kitchen table in the Netherlands, in this chapter, I traced some of the myriad relationships that constitute a landscape as identified through beansites. The four stages of CHORA's Urban Gallery avoid privileging human over nonhuman ecologies, or vice versa, and a design process

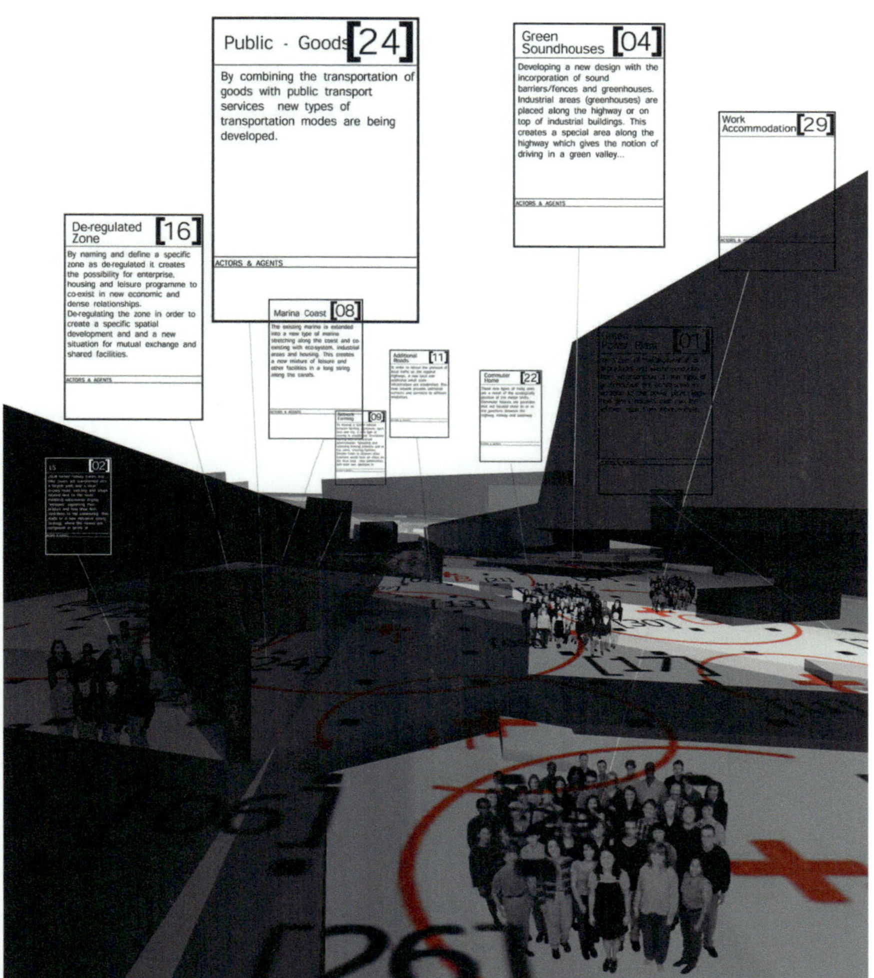

Figure 43. Sector E, Project W. Dynamic masterplan consisting of protoypes. (CHORA)

emerges that is deeply rooted in the operations of a landscape. We amassed operational fields by conducting miniscenarios; we combined these operational fields to create prototypes, which we then tested by playing scenario games. Games bring together multiple perspectives and challenge hierarchies. Should the prototypes survive the extreme scenarios of the game, they are then developed into actionable proposals in an action plan.

That a project may have a prototypical impact on a wider landscape is an anthropological way of approaching design. Anthropologists unearth the relationships, obvious and latent, between an object and its context—whether that context is environmental, physical, political, or social. To this end, one might begin to understand a prototype as the material embodiment of anthropological data gathered from fieldwork. One might regard the operational fields as the ecologies, in a Guattarian sense, that constitute a landscape whether these ecologies be environmental, social, or subjective, or otherwise. Bunschoten, and others, describe a form of landscape that does not privilege the urban over the rural, nature over culture, spatial over the emotional, human over the nonhuman, dreams over reality, science over feeling, this world over the next. The CHORA process reveals the concept of a landscape composed of an assemblage of operational fields. As Clifford Geertz reminds us, "Man is an animal suspended in webs of significance he himself has spun."[60] The study of these webs of significance is an interpretive study looking for meaning, rather than objective facts, *per se.*

To understand this relational landscape, we must understand the operational fields at play. To understand landscape, we need to trace the associations; to design the landscape, we need to reassemble the associations. What is landscape, if it is not the construction of relationships? Understanding and designing go hand in hand. Both are deeply rooted in the knowledge and insights gained from landscape fieldwork and, one might add, the generative potential of beansites.

3

WALKING AND TALKING

Chromatic Metaphors in the Bahraini Landscape

Three things take away sadness: water, greenery, and a beautiful face.

—Arabic saying

I first heard the Arabic word *shlawnak* as my friend Isa drove me home to my lodgings after dinner at his parents' house. It was shortly after my arrival in Bahrain for a year of fieldwork. As we made the brief trip across the island that evening, I noticed that every time Isa answered his mobile phone, he would say, "Shlawnak?" raising the pitch slightly toward the end of the sentence as if posing a question. At first, I thought Shlawnak was a friend of Isa's who kept calling him. However, when I asked Isa who Shlawnak was, he explained that it was a colloquial greeting. Literally, it means "What's your color?" Used as pervasively as "How are you?" or "How's it going?" in the Arabian Peninsula, the question implies that color is a measure of a person's well-being. "Shlawnak?" invariably brings the reply "Al-ḥamdu l-illāh," "Praise be to God," rather than the color the question literally asks for.

The lines above are adapted from the closing paragraph of my book *Paradoxes of Green: Landscapes of a City-State*—an ethnography of green in Bahrain.[1] I was unaware of the Arabic word *shlawnak* until my conversation with Isa that evening, having not encountered the *Khaleeji* (meaning Gulf) Arabic slang word in my modern standard Arabic lessons.[2] Indeed, I would probably never have come across *shlawnak* had I not spent an extended time living in Bahrain. Meanwhile, Isa was amused by my excitement when I learned of the word's meaning. Despite his knowing more than anyone else in Bahrain what I was doing there, it had never crossed Isa's mind to tell me about *shlawnak*. It was only through my immersion in Isa's everyday life practices—in this case, dinner with his parents and a ride home—that I was introduced to *shlawnak* and what it might signal about the pervasiveness of color in everyday life.

Later, I began to speculate on the significance of the word and its relevance for my research on chromatic concepts. The ethnographer often works with seemingly small details, such as a single word, that lead to larger conclusions and generalizations.[3] Isa's use of *shlawnak* had confirmed one of the main takeaways of my ongoing research; the color green, I was finding, was a measure of human health and of a landscape's health too.

Fieldwork Design

Isa was one of several close friends and interlocutors I spent time with during my year of ethnographic fieldwork in Bahrain as part of a doctoral degree program at Harvard University, under the supervision of Hashim Sarkis. This was after spending a few years based in London as a practicing landscape architect and part-time tutor. I went back to graduate school, in part because I wanted to have more dedicated time to devote to a deeper understanding of how people come to know landscapes. Isa and I would regularly meet in the late afternoon or early evening after the sun went down and after he finished his work in the civil service. We would have dinner together and explore various parts of the island in his car and on foot. Sometimes we met on Fridays for lunch with his parents who told me to treat their house as my home. We talked about many things, including Bahrain's physical, political, and social landscapes, but also many topics outside the scope of my research. I spent most of my days in that way—talking with people, using a technique of unstructured interviews, a time-consuming method designed to let conversation flow, while all the time carefully crafting probing questions so as not to prejudice the answers. While I knew many Bahrainis before my visit, I wanted to meet people outside my friends' social orbits too. I met countless new people through the practice of walking. Chance became a big part of my method, and walking increased the chance of chance encounters.

As I walked through Bahrain's landscapes and recorded my encounters with green, I learned local codes, dialect, and patterns of behavior. When I would meet green—either as color or as an environmental movement—I would pause to talk with people about its significances as well as write, photograph, paint, and sketch my impressions. Sometimes I would take scratch notes should something come up in conversation that could merit further thought or investigation. When I got home in the evening, I would write up my fieldnotes. The anthropologist Roger Sanjek describes a third type of note, the headnote. A headnote is retained in the mind. Sanjek points out that ethnography is the product of fieldnotes and headnotes and that headnotes ought to take precedence over fieldnotes.[4] But why? Well, one of the reasons relates to the interpretative dimensions of fieldwork. As I read my fieldnotes now, fifteen years later, my memory of events often differs from the way they are described in my fieldnotes. This can be because my memory has shifted, or with the benefit of hindsight, the interpretation of the event is different, and it takes that time to interpret events. With Sanjek's advice, my headnotes take precedence over what I wrote at the time.[5] When I returned to the United States after my year of fieldwork, I began to write up my ethnography, returning to Bahrain periodically to check facts and recalibrate myself. While writing, I returned to reflect on the meanings and significances of my fieldnotes. The entire iterative process took ten years from fieldwork to book, which is not unusual for ethnographic fieldwork.

What became the book *Paradoxes of Green: Landscapes of a City-State* is based on an ethnographic account of the year I spent in Bahrain, walking through its landscape, talking to people, and interpreting the various values attached to landscape and to greenery. One of the book's contributions to knowledge is understanding landscape through the lens of color. Color is an essential component of design, yet designers rarely discuss color for some reason.[6] Through my fieldwork, I came to understand that color and landscape are deeply ingrained in Bahraini everyday life and, in fact, are mutually defining.

My fieldwork exposed several paradoxes relating to green and greenery. There is often a general assumption, at least among landscape architects, that landscape architecture is good for cities because of its greenness. This equating of green with environmentalism is mostly a Western construct, originating in the verdant landscapes of temperate West Germany in the 1970s.[7] But "green" does not always mean sustainable. To create greenery in a dry, arid desert landscape such as Bahrain's is not very green in an environmental sense, due to the amount of water, chemicals, and labor resources required to maintain the greenery. As I learned from fieldwork, over half of Bahrain's water, which today comes almost entirely from desalination processes, is used to create greenery. After water is desalinated in Bahrain, most of it is then poured onto the land to create value. Through a focus on green, my fieldwork demonstrated that landscape is not just a central object of intervention for the political, economic, and social life of Bahrain as a city-state and nation-state, but that landscape, in turn, profoundly shapes social, economic, and political life.

Notes on Process

The year I spent living in Bahrain, in intimate contact with landscape and people—or, more accurately, "landscape," since landscape already includes people—led me to knowledge that would have been difficult if not impossible to achieve in other ways. The extended period in the field also led to a very different engagement with my interlocutors than would have been the case otherwise. Still, the year in Bahrain was certainly not easy. Insightful encounters, such as the *shlawnak* moment, were punctuated by long periods of stagnation when seemingly nothing happened.

One overarching question I address in this chapter is: Was the ethnographic fieldwork worth the investment? For sure, it took a lot of time, effort, and money, but was it worth it? In the pages that follow I discuss some of the value of ethnographic fieldwork for landscape architects and others interested in the design and construction of the built environment through a behind-the-scenes view of that year in Bahrain. How did the in-person fieldwork lead me to insights I wouldn't have reached through researching in a library or an archive or talking to someone over the phone or online?

In reviewing what I did, first I will first address two questions: Why Bahrain?

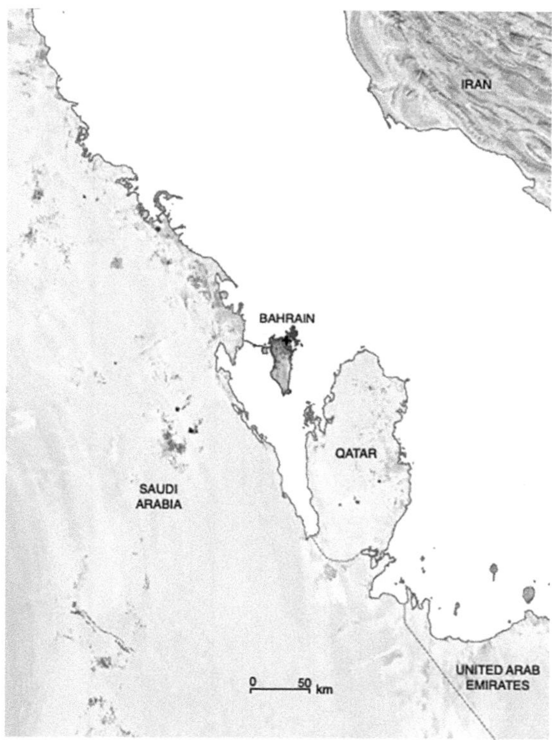

Figure 44. Bahrain is located twenty miles (thirty-rwo kilometers) from the coast of Saudi Arabia. (Map by Maria Vollas)

Why green? Then, moving through the four above phases, I will discuss the methods I developed while conducting fieldwork and writing the book. I will conclude by outlining some of the strengths and limitations of doing landscape fieldwork, and share some lingering questions, design implications, and concerns that have emerged since the publication of the research in book form.

Why Bahrain?

Before explaining why I chose to conduct fieldwork in Bahrain, allow me to set the scene with some context on the place and its history. The Kingdom of Bahrain

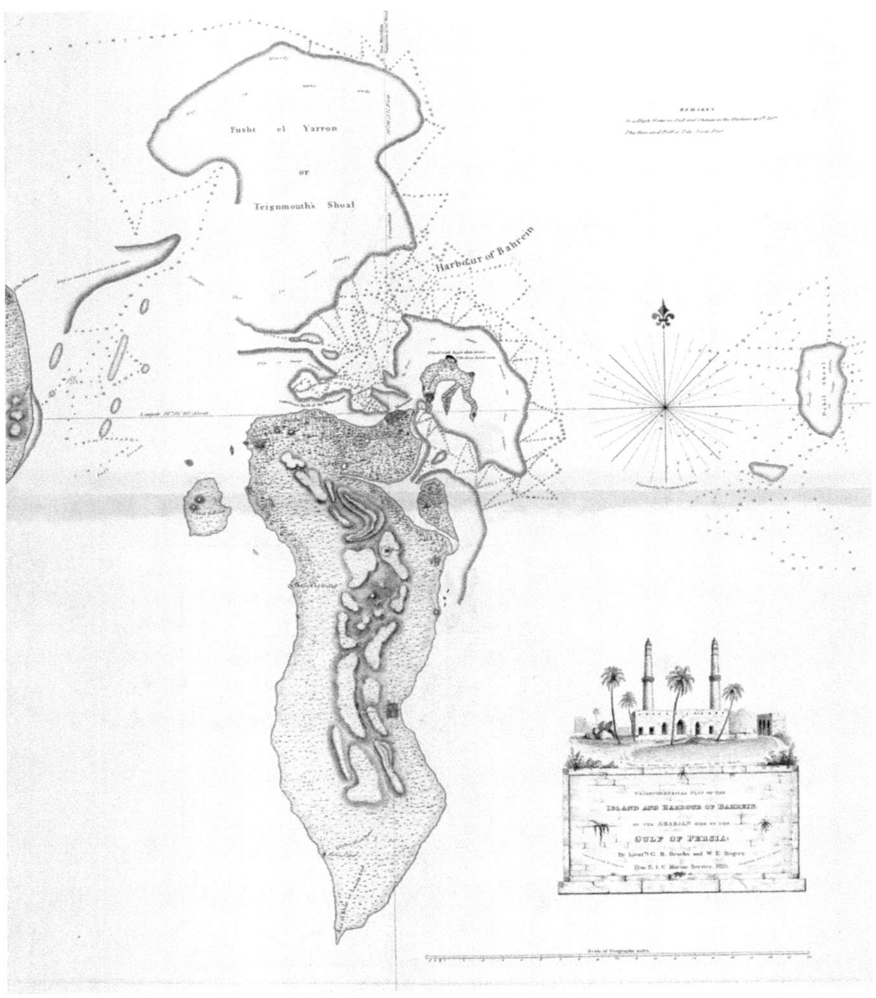

Figure 45. "Trigonometrical Plan of the Island and Harbour of Bahrein, on the Arabian Side of the Gulf of Persia; by Lieutts. G. B. Brucks, and W. E. Rogers. Hon. E.I.C. Marine Service: 1825. Drawn by Lieutt. M. Houghton; Draughtsman Hon. E.I.C. Marine. Engraved by Bateman and Son" [21r] (1/2). (British Library: Map Collections, IOR/X/3630/21, in Qatar Digital Library, accessed July 14, 2023, https://www.qdl.qa/archive/81055/vdc_100023868004.0x00002a)

is small and self-contained. Only ten miles wide and thirty miles long (sixteen by forty-eight kilometers), Bahrain is an entire nation-state in one walkable area. At just over 760 square kilometers, or roughly 300 square miles,[8] the kingdom's total area is just larger than Singapore, and just smaller than New York City. An archipelago of thirty-three islands, most of the population is centered on the main island, Bahrain, and the smaller island of Muharraq. Bahrain is the smallest, densest, and greenest state in the Arabian Peninsula.

In terms of climate, Bahrain has effectively two seasons: a very hot summer, and a moderate winter. Temperate plant species, such as roses, prefer the winter months. In summer, temperatures regularly reach above 40 degrees Celsius (104 degrees Fahrenheit) and there is very little rain. The annual rainfall, which is concentrated in the winter months, is a paltry 81.65 millimeters or 3.2 inches, compared with 1,169.07 millimeters or 46 inches in the United Kingdom, and 741.13 millimeters or 29 inches average for the United States.[9] Bahrain is so arid, and the rains so infrequent, that there are no permanent rivers or streams on any of the islands. However, springs punctuate the seabed as well as the land of the island. Residents have relied on about twenty springs offshore and fifteen springs on land to sustain agriculture in Bahrain for millennia. The waters give Bahrain its name, which literally means "Two Seas" in Arabic—the first being the Gulf, and the second sea being the underwater and underground springs.

From 1861 until gaining independence in 1971, Bahrain was a British protected territory, or trucial territory, under indirect British colonial authority called the Persian Gulf Residency. Descriptions of Bahrain in the early 1900s bear little resemblance to the islands of today. A British colonial report from 1905 described Bahrain as "a pleasant oasis," then goes on to describe: "The golden dusted roads which cross it are broad and shaded on either side by long forests of date palms, deepening into an impenetrable greenness, cool with the sound of wind among the great leaves and the tinkle of flowing water."[10] The flowing water refers to the *aflaj,* or water channels, that used to run through the date palm groves for irrigation.

Unforeseen events disrupted the pleasant oasis and tinkle of flowing water and changed the islands forever. The main driver was discovering oil in the 1930s—making Bahrain the first of the Gulf countries to do so—which irrevocably altered the social and political ecosystem of the islands. The consequent increase in incomes, healthcare, life expectancy, combined with immigration associated with the economic and social transformation following independence in the early 1970s, resulted in an explosive increase in population. In the 1920s Bahrain's population was just 70,000; in 1971, when Bahrain gained independence from the British, the population exceeded 200,000. In 2006 the population was estimated to be 700,000 (including 250,000 foreigners, mostly from India and Pakistan, and around 3,000 British citizens). In 2008, the population passed 1 million, and in 2024, the population had exceeded 1.8 million.[11]

Sources in Bahrain tell me that the countryside and city have been separated for at least a couple of centuries, since the arrival of the ruling Al-Khalifa family in 1783, but this division has became starker since independence in 1971. Independence coincided with the arrival of city planning in Bahrain. The first planner, A. M. Munroe, was appointed by the newly independent state and set about developing masterplans for Bahrain. These plans delineated city and country as separate entities with a green belt around Manama.

Around 70 percent of the current Bahraini population is Shiʿi and 30 percent Sunni—referring to the Muslim sects—though it must be fully acknowledged that such categories are woefully inadequate to explain the complex ways people understand these labels socially, politically, and religiously. In the mid-twentieth century, the Shiʿi and Sunni population ratio was more balanced, and the demographics are again levelling out as citizenship is given by the state to Sunnis from other Arab states. Given that the rulers of Bahrain are by-and-large Sunni, these demographic divisions and shifts often lead to political, social, and cultural tensions. Writing about the nineteenth century, the urban historian Nelida Fuccaro describes Bahrain as having had a three-tier social hierarchy: tribal Arabs, who were Sunni; the Baharna, who were Shiʿi and usually lived in villages in or near the date palm groves on the north and west coasts; and the merchant classes, which were either ʿAjam or Hawala. Others point out that these categories and hierarchy do not do justice to Bahrain's complex sociology of the nineteenth century, and certainly not today. Today, this mix is complemented by expatriates from the Indian subcontinent; nearby Arab states including Jordan, Lebanon, and Syria; and expatriates from largely professional classes from the United States, Australia, New Zealand, and the United Kingdom and Ireland.[12]

When I first traveled to Bahrain, it was at the invitation of the Bahrain Society of Engineers, who invited me to teach a three-day course on landscape architecture. This came about as the result of an advertisement in *Landscape Design,* the journal of the Landscape Institute in London, where I then lived. The course was inspired by the work of CHORA, described in chapter 2. As I explored the islands for the first time on my short three-day long visit, I was astonished and intrigued to see so much green space in such an arid, dry, desert landscape. On my first evening, I was driven through a date palm grove on the west coast by a sheikh from the ruling family, a cousin of the king. On a subsequent evening, I walked through the streets of the capital, Manama, during a sandstorm. I had never experienced such an all-encompassing beige atmosphere before. Amplified by the orange sodium streetlights, the beige landscape was not just on the ground but in the air too. It reminded me of the importance of the desert across faiths as a place where we can hear God speak: the hostile desert focusing minds on the essentials of the human condition and providing a link to the transcendent. I described this walk in my fieldnotes as "a spiritual experience."

Why, I wondered, were there so many attempts to build greenery in this inhospitable climate? Each time I visited Bahrain to teach more classes—and the BSE invited me back annually—I became more and more engaged in this strange but familiar landscape. The strangeness came from the aridity, the dryness, and the beigeness, while the familiarity, I realized, came from Bahrain's postcolonial condition as a former British-governed territory: the road markings, the English street signs, the Cadbury's chocolate all felt familiar. The greenness of Bahrain, meanwhile, was both strange and familiar, depending on the hue.

When it came time for me to choose a field site for my doctoral research, I was attracted to Bahrain in part because I imagined that I could more easily and precisely trace the associations between hard and soft infrastructures in this tiny city-state and island nation, as opposed to a larger city with a more expansive hinterland.[13] But I could also help in some small way to decolonize the Western-centric literature. There were also personal reasons why I chose Bahrain. The Inishowen peninsula where I grew up in Ireland has the same land area as Bahrain, with a proximate city of Derry-Londonderry, just like Manama in Bahrain (see chapter 1). This made the strange landscape seem more familiar to me, while providing a basis for cross-cultural comparisons.

I was fascinated by Bahrain as an "extreme case" for the questions I wanted to explore. Albert J. Mills, Gabrielle Durepos, and Elden Wiebe, all experts in organizational management tell us in their *Encyclopedia of Case Study Research:* "The extreme cases approach is employed when the purpose is to try to highlight the most unusual variation in the phenomena under investigation, rather than trying to tell something typical or average about the population in question."[14] The assumption is that by looking closely at unique variations, we can then be better prepared not just for the actual extreme situations—in this case, concepts of green in arid desert landscapes—but, importantly, more temperate climates or moderate cases too. Bahrain's complexities and extreme environmental, social, and political conditions contributes to the obsession for the creation of urban greenery. However, this compulsion is not unique to Bahrain; it's a global phenomenon that happens to be stronger in Bahrain. For all these reasons, Bahrain seemed like a fitting place to explore the questions I had around color, landscape, and urbanism.

Why Green?

In the beginning of my fieldwork, I set out in search of the "urbanism of landscape." By "urbanism," I meant the complex hard and soft infrastructures—or the physical as well as the sociocultural and economic infrastructures—that sustain landscape in Bahrain. I was also interested in the relationships between people and the land in this harsh desert environment. I noticed a pattern: new building developments were preceded by the designed landscape. This intrigued me. In the West, the landscape is often relegated as the backdrop for architecture; it's the last part of the design to be built, depending on the budget available. But at the time I was reading literature on landscape urbanism, a disciplinary movement based on the idea that landscape, rather than architecture, is the basic building block of urbanism. In other words, the city is made up of an accumulation of outdoor spaces, rather than buildings.[15] In Bahrain, I noticed that developers advertised buildings based on the promised landscape, and not the architecture. I seemed to have happened upon evidence for landscape urbanism, or landscape architecture that preceded architecture, and I wanted to know more about this condition. So, I set out to find out why developers were constructing the landscape before they built the houses.

First, though, I had to figure out what most people meant by "landscape." In the classroom on my first visit to Bahrain, which was offered in a conference room in the Gulf Hotel, I had become aware that my definition and those of the students did not line up. For the course's participants—mostly Bahrainis, but also some Emiratis, Kuwaitis, Saudis, and Omanis—landscape had a limited definition: a green garden. Meanwhile, I was interested in larger landscapes, "the object spread out beneath an airplane window,"[16] and beyond, which might include the desert, cities, and human settlements, as well as green space. This disconnect continued throughout my fieldwork; whenever I tried talking with Bahrainis about landscape—no matter what form of landscape I tried to discuss—they would regularly end up mentioning green. In fact, my interlocutors understood spaces as "landscape" only if they were green—a color often mentioned in Bahraini poetry. My interlocutors were especially interested in the green gardens of the elites, or date palm groves, or other luscious green spaces. I came to understand that for Bahrainis, the creation of green is the biggest change that can be made to the indigenous desert floor. My inkling that green would be especially important was supported by a beautiful Arabic phrase—which is often misattributed to the *hadith* but my research suggests is pre-Islamic—that illustrates the importance of green in that part of the world: "Three things take away sadness: water, greenery, and a beautiful face" (*Thalāthat ashyāʿ yudhhibn al-ḥuzn: al-māʿ wa-l-khudra wa-l-wajh al-ḥasan*).[17]

In *Paradoxes of Green,* I write that I focused on green because that was my interlocutors' category for landscape. It is true that Bahrainis had different understandings of the word landscape than I did, and that in conversation with me they would invariably associate landscape with green. So, as a result, I focused my research on green, rather than on landscape *per se.* On reflection, I see that while that was the case, it was not that simple: I freely admit that I may have brought my own assumptions to the equation too. With the benefit of hindsight, I recognize that my interest in color precedes my time in Bahrain and may have influenced my thinking at the time. Tellingly, my fascination with Roberto Burle Marx, who had inspired me during my early research and design projects as an undergraduate student and continued to motivate my work in Bahrain, was in part due to his affecting ruminations on the interrelationships of landscape, color, volume, texture, and form. While some may see such a predisposition as a weakness of ethnography as a method,[18] I do not

Figure 46. Bahrain land cover change, 1987–2023. (Maria Vollas)

see it that way; if we learn to recognize the biases of the ethnographer and work with them, we may turn this to our advantage. In this case, I approached Bahrain's landscape from the perspective of a landscape architect interested in color. This landscape architecture "bias" allowed me to focus and instrumentalize my research in a way I might not otherwise have been able.

Stages of Fieldwork

Early in the project, I realized that because there were few direct answers to the sorts of questions I was asking, if I wanted to work on Bahrain then I needed to spend a lengthy period living there. But my landscape fieldwork did not start when I arrived in Bahrain and end when I departed a year later. The several visits I made to Bahrain before and after my extended fieldwork all count as fieldwork. While I may not have kept notes to the extent I did when I was "in the field," these early and later visits were pivotal. The earlier visits piqued my interest and gave me something to be intrigued about. I learned that I could not count on received data; the available information was limited and unreliable in Bahrain, where there are no public information laws. I needed to gather my own data from multiple sources. In the end, I pieced data together from various sources, including people I encountered during my walks and talks. My later visits to Bahrain allowed me to question the observed knowledge and to double check the data received in the field, aiding in my interpretation and analysis. With the benefit of hindsight, I can delineate four distinct phases to my fieldwork in Bahrain:

1. Remote and preliminary fieldwork
2. Acclimatization
3. Becoming
4. Writing up and fact-checking

Bruce Jackson, a folklorist known for his work on penitentiaries in the American South, describes three phases to fieldwork: planning, collecting-in, and analyzing.[19] While Jackson recognizes "collecting-in" as one phase, I am splitting it into two distinct but overlapping phases: acclimatization and becoming.

Remote and Preliminary Fieldwork

As I readied myself for fieldwork while still in the United States, I met with Bahrainis as often as I could and talked with them about Bahrain. I had met many of these friends through my visits to teach courses for the Bahrain Society of Engineers, which continued for three years after my first visit. These courses gave me the opportunity to visit and explore the Bahraini islands prior to fieldwork, as well as meet many Bahrainis and other professionals from around the Gulf. With ten to fifty students enrolled in each training course, I met people from different segments of the community who represented different social groups and religious and political persuasions. I also met Bahrainis who worked for government ministries and other public institutions focused on improving the built environment, which I hoped would help with access to information and the gathering of data. I turned out to be wrong on that front, but nevertheless I learned from that experience, and my friends were as generous and helpful as they could be.

Before departing for the field, I also regularly visited the islands on Google Earth and familiarized myself with their landscapes as best I could from a computer screen. I presented my work at venues such as the Gulf studies conference at the University of Exeter's Centre for Gulf Studies, a workshop on Port Cities at the Harvard Kennedy School's Middle East Initiative, The Right to Landscape conference at the University of Cambridge's Jesus College, and an invited presentation for the Charles McGlinchey Summer School in Ireland. In doing so, I met other researchers of the Arabian Peninsula or who were involved in similar fieldwork, including Thomas Brandt Fibiger, Nelida Fuccaro, Ahmed Kanna, Sawsan Karimi, James Onley, Martin Hvidt, and Hussain Ilias.[20] I continued attending and participating in other conferences during my fieldwork, which proved an efficient way to meet others with the same interests. Later, I had the good fortune of overlapping with Nelida Fuccaro in the field. Nelida's book *Histories of City and State in the Persian Gulf: Manama since 1800* is an essential history of the development of Manama and Bahrain.[21] Nelida's in-the-field insights and introductions to her interlocutors, who also became my interlocutors, supplemented and transcended what I learned from reading her academic text.

A fieldworker's first loyalties are to a field site, but they are situated within a field of studies as well. In preparation for my fieldwork, I took courses in Arabic language and on Islam and the Middle East. I also read everything I could find about Bahrain, getting to know the islands' histories as well as their geographies and demographics. A key primary text was Charles Dalrymple Belgrave's *Personal Column,* an autobiography by the former long-term advisor to the emir of Bahrain, the head of state.[22] In this role, between 1926 and 1957, Belgrave was effectively prime minister. He painted and kept a detailed diary which eventually informed his book that documented the development of the islands during the years following the discovery of oil and eventually leading up to independence.[23] I spent some time in Belgrave's archive at the University of Exeter, which contains copies of his diaries as well as his paintings. I was struck by the absence of green in his paintings, not to mention the callousness of some of the comments in the diary. He mentioned on one occasion how the police were too cowardly to carry out an execution, so he stepped in and finished the deed, then went home for breakfast.

Books about Bahrain's landscapes are few and far between, with just a few references on the physical and botanical life of the islands.[24] There are some studies on the social and political history of the islands,[25] but very little has been published on architecture, landscape architecture, or the architecture of the landscape.[26] At the time I was deeply motivated by writings of Kenneth Frampton, Ian McHarg, Jala Makhzoumi, Mohsen Mostafavi, Dan Rose, Hashim Sarkis, and Anne Whiston Spirn, Charles Waldhiem, and others. I was also reading ethnographies of the Middle East, building upon my previous forays in ethnography at the University of Pennsylvania. I later came to understand that while many of these authors were effectively writing about the same thing, their work came from different genealogies. This is illustrated by the two books I brought with me to the field: the anthropologist Steven Caton's *Yemen Chronicle,* a so-called ethnomemoir or reflection on his fieldwork in Yemen conducted twenty years before; and *The Pine Barrens* by the writer John McPhee, a wonderful "thick description" of a New Jersey forest that describes the interactions of people and the land. All of this background knowledge helped prepare me—at least somewhat—for the field.

Acclimatization: Ministry Work

After arriving in Bahrain, I found a place to live in Gufool, about thirty minutes' walk from the center of the capital, Manama. I had arrived during Eid, which marks the end of the Ramadan fast. I had been told to expect to be invited to different family gatherings by my various Bahraini contacts. No one invited me: it was one of my first glimpses at how difficult it is for a foreigner to penetrate Bahraini life. Nevertheless, I kept myself busy looking for accommodation in the Internet cafes of Manama. Eventually, I found a small apartment through a website called expatriates.com. What stood out to me in the ad was the description that it is in a "nice green area."

The first months in Bahrain were the hardest, as I adapted to my new environment. Thankfully, I had secured an informal but regular position at the Ministry of Municipalities and Agriculture Affairs. Although my appointment was only officially approved after my fieldwork had ended, I was given a desk in the ministry's office at Gold City in the suq, or marketplace, the term given to the older part of Manama. This was the secondary building across the street from the main building on Government Avenue, where the minister was based.

After I got settled in, on a typical day, I would rise in the early morning, often woken by my landlord working in his hair salon. After breakfast I would write, sometimes working on papers, conference presentations, and the like. If it was a weekday, I would walk to the ministry in the suq. I tried to keep ministry hours, arriving around 7:30 am and staying till 2 pm. After lunch in the suq, invariably

Figure 47. Gufool, Mamama, Bahrain, with a patchwork of square blocks in the center and water gardens to the right of center. (Google Earth)

green pea masala—extra spicy—in the former Nirvana Indian restaurant, I would return to work for a little while before going home for a nap, like many of my colleagues in the ministry.

I spent my days in the ministry touring different offices and getting to know different officials, especially those people who had interest and influence over the built environment. In some ways, perhaps, on reflection I was recruiting allies in some ways. I set up meetings with those officials who were recommended to me. Others, I met on the hallway or in the elevator, and these chance encounters often led to further exchanges over a tea or coffee. Among those I met was Dr. Youssef, an advisor to the minister. Dr. Youssef was an admirer of Patrick Abercrombie, the founder of the postwar London greenbelt and the new towns movement.[27] Abercrombie's "potato plans" represented cities from the perspective of agglomerations of neighborhoods and functionalities. Abercrombie's ideas had a huge impact on the physical environment of Bahrain, including the greenbelt that was being constructed in Bahrain.[28]

In those few early weeks, I reached out to many people. In addition to meeting as many people as possible through the ministries, I joined groups and gave lectures at the University of Bahrain. Initially, I was discouraged when people did not show up for prearranged meetings or canceled at the last minute. They often said it was "not God's will" that we should meet on that day, but we'd meet another day. I suspect there was an element of testing going on: They were seeing how serious I was. When I persisted, the meeting would generally happen. I came to meet several of my interlocutors through sheer tenacity. Only once did a Bahraini reach out to me first. Dr. Juma found me through a mutual contact and requested that we meet at Bubbles Café in Salmaniya, not far from my lodgings. The meeting with Dr. Juma was significant because he wanted me to interview him, reversing the usual dynamics.

In my notes from those early weeks, I mostly wrote about the mundane aspects of my daily existence in Bahrain. Admittedly, there were days when I did not write. I felt there was nothing going on, and I was too despondent to record what I thought was my noneventful life. But this was a mistake. Over these few weeks, people were assessing my commitment to the field. Later, I tried to reconstruct these lacunae by analyzing the emails I sent to friends and family. Combining my personal communications with photographs and sketches, I was able to get a better sense of what happened in those days. I learned an important lesson: Even when we think nothing is happening, the world is still moving. It was here that I learned of the power of intermediality, and the juxtaposition of different media, text and image. One informed the other, creating something else.[29]

Acclimatization: Walking as a Method

With the setting of the sun, the *adhān,* or call to prayer, would announce the arrival of the evening. Then, I would walk. Sometimes to prearranged meetings, sometimes just walking aimlessly, but it was then that I got to know Bahrain best. Most of my fieldwork in that way was done at night. Nighttime is also when public spaces come alive.

Mostly I walked alone, but I also walked with the Bahrain walkers, an informal group, which often led to intriguing conversations. Indeed, Albena Yaneva discusses walking as an interview method in her book *Crafting History: Archiving and the Quest for Architectural Legacy,* where the act of walking facilitates the conversation.[30] Although I did not really consider walking as a method before I traveled to Bahrain, it is fundamental to my understanding of Bahrain. I found that walking could complement rather than replace satellite images, which were also an important research tool, as I used Google Earth satellite images to choose my routes. In other words, there was interplay between the aerial image and what I saw as I walked. In my attempt to describe the urbanism of landscape, I was motivated by an observation by Charles Waldheim: "New audiences and sites for work also offer the possibility of new formulations of landscape, recasting its image from green scenery beheld vertically to a flatbed infrastructure that includes both natural and urban environments."[31] Waldheim's challenge to change our perspective on landscape toward a horizontal surface that makes no distinction between landscape and urbanism, privileges the aerial over the eye level. Why not behold landscape from both perspectives, from above and from eye level too? Similarly, as Walter Benjamin reminds us, "The power of a country road when one is walking along it is different from the power it has when one is flying over it by airplane."[32] Benjamin goes on to recommend: "Only he who walks the road on foot learns of the power it commands."[33] Even though the aerial image can show how a landscape is arranged formally, we can only really know a landscape and how it is inhabited from the inside. Francesco Careri, founder of the Italian urban art workshop Stalker (*Osservatorio Nomade*), describes walking "as a primary act in the symbolic transformation of the territory."[34] For Careri, the act of walking is an intervention that changes the landscape and perceptions of it.

An essential aspect of such walking practice was that the walks would introduce me to people I would otherwise not have met, and I would talk with them about green. I walked in order to talk. In the year of my fieldwork in Bahrain, and inspired by Richard Long, the land artist, I initially toyed with the idea of drawing straight lines on a map and walking those lines as a sort of performance as well as a recording and documentation of green. Since we rarely experience the landscape through a perfectly straight line, the aim was to construct a whole new reading of Bahrain. I also considered a random sampling method, based on my experience working at CHORA in London. This method too would have brought me to places I might not otherwise have visited, not to mention to meet people I might not otherwise have met. In the end, though, I wanted to focus on green, through its presence in the landscape and in the imaginaries of the inhabitants of the landscape. I had to follow their concepts.

Using aerial images coupled with intuition, I took a pragmatic approach to my walks, deciding on what I imagined as the most interesting route, but not too long either. I walked everywhere I needed to go. If I was going to a ministry in Manama to discuss green, I would walk through the suq to get to the ministry taking a slightly different route each time. The narrow streets of the suq were populated differently depending on the time of day or night. Mostly, they were populated with low-income workers from the Indian subcontinent who lived there in densely packed accommodations. I recorded my interactions with green along the way, taking photographs and writing scratch notes in my pocket-size notebook and methodically noting details not just of the green but of what I understood as people's reactions to green. When possible, I would start conversations with strangers or at least be receptive to strangers starting conversations with me. Being densely built, the suq is not very green, but still I would note the green vegetables being sold, the green winter *thiyāb* for sale in the tailor's, the green hubs on the car parked along the sidewalk, the green weeds poking through the pavement, the *mashmūm* (sweet basil) for sale at

Figure 48. Sh. Ibrahim's Garden, 1963. (Photograph by Torkil Funder, courtesy of the Moesgaard Museum)

the cemetery on a Thursday evening. If I had to go farther than I could walk, I would either take public transport, or Illias, my landlord's driver, would drive me. Rarely did my routes turn out as expected—and that was the point.

Although appearances can be deceiving, and there are many Western-looking people who grew up in Bahrain, it's fair to say I do not look like a typical Bahraini. As I walked through the suq on my way to meetings, I inevitably stood out. I often dressed in business attire during the daytime, with a suit and tie, or at least a shirt and tie, which made me stand out even more; not many other Western people would walk through the suq wearing a suit and tie, but business dress tends toward the formal in Bahrain and for meetings in the ministries I erred on the side of caution. In some ways, the clothing made it easier for people to speak with me. They wondered if I was lost, and if they could help me. Eye contact was also important—once I took the initiative, people talked. Occasionally, I would carry my camera in my hand, making me look even more of a foreigner. Tourists are not that plentiful in Bahrain, so the assumption usually was that I was some sort of expatriate professional working for the government.

At nighttime, I would wear more casual clothes, maybe jeans and a shirt, just like my Bahraini friends. However, I still managed to stand out. Random conversations with strangers—mostly men—were more difficult at night when one's motives are more likely to be misconstrued. Through walking, I became aware of other illicit, secret, and symbolic uses of public spaces. For instance, there were several instances of individual men approaching me in public spaces looking for something more than friendship. Bahrain has gay spaces, just like other cities, although they are perhaps less obvious. As a man walking alone, it was often assumed I was looking for companionship. In general, being a man, and a foreign man at that, offered me many opportunities that would have been more difficult for others. The people I would meet at random were generally men. Men would talk to me because I was a man. As a man, I seemed to have been granted more right to public space than a woman would have been in Bahrain.

Becoming

There were two parts to the becoming phase during my fieldwork: Bahrain accepting me, and me accepting Bahrain. This phase was full of turning points where I felt less like an outsider looking in, and more like an accepted insider looking out. The becoming process unfolded through a series of events and happenings over the course of the year, some small and private, and others more public and external. The first of these private moments of induction was when Umm Isa, my friend's mother, told me that their house was my home. This gave me a base, a safe space to return to. Such moments were essential for developing an intimate relationship with my interlocutors.

One of the most important moments in my fieldwork took place in a dilapidated date palm grove that Isa and I nicknamed the "Secret Garden." Isa and I visited the grove very early one Friday morning. Isa had insisted we see this forgotten grove, adjacent to the sea, that harked back to a bygone era. As we walked through the date palms, and Isa described the long history of the space, I felt sad to hear of the likely impending destruction of the grove. There, I had an epiphany of sorts. I came to understand that color has memory. The melancholic greens we were seeing had been built up over millennia; they would not be easily replaced. Yet these greens were being lost all over Bahrain.

I had previously gotten into an argument with a developer I met, Nabil, who contended that one can just cut down the date palms, then build houses, then plant more palms, and that it would look the same. "But it is not that easy!" I tried to explain. There is more to green than meets the eye. The color is built up over millennia of landscape practice and can't be easily reproduced. Nevertheless, Nabil and

Figure 49. Date palms, Saar, 2022. (Photograph by Camille Zakharia)

his colleagues had proceeded with removing date palms and planting villas in the groves, often with green-painted roofs to compensate for the date palms that had been displaced. The hues of the green roofs, though, were brighter in hue than the date palms. These preferences reflected the changes in values attached to greenery since the discovery of oil in the 1940s, and the consequent industrialization of the country. Nowadays, the gray-greens of the date palm plantations are not valued as they once were. Farmers who still practiced agriculture did so not primarily for profit or need—it's much more economical to import—but for aesthetic and emotional reasons. But most residents wanted grass-green lawns and bright green roadside verges and golf courses—all of which took a tremendous amount of water to sustain. Finally, I understood why new developments were being sold on the greenness of their surroundings, which were never as green as the images would lead us to believe.

Another significant—and more public—moment for me came when I participated in the annual ritual of ʿAshura', a holy day which both Shiʿa and Sunni Muslims observe but in different ways. In *Paradoxes of Green,* I describe participating in some Shiʿa events surrounding this holiday, where I was splashed with human blood during a densely crowded procession through the capital:

> I spent an hour or so winding through the shaded narrow streets of Manama, taking photos and being drawn into the euphoria of the event, the rhythm, the chants, the beating, and the blood. It was likely someone hitting their head in order to draw more blood that caused my face to get sprayed with blood. As I sensed the drops running down my face, in one moment my enthusiasm left me, as if every drop of my own blood had drained from my body. Suddenly forlorn, disgusted, frightened, sickened, I withdrew from the crowd and found somewhere to sit on the pavement.[35]

The above event was significant for two reasons. The first, as I describe, was the physical anointing with blood. This was a rite of passage that helped me become a Bahraini in my eyes and in the eyes of (some of) my interlocutors. One of the main consequences of my experience with ʿAshura' was that it introduced me to many Shiʿa, the majority population who until then had been outside my social orbit. I came to realize that my association with the Sunni-populated Ministry of Municipalities and Agriculture Affairs was not all that helpful for my research. I started to spend less time in the ministry and more time on the streets and roads, walking and talking with strangers. The event was also significant on the level of embodied stimulus. To this day, I can vividly recall the smells, the sounds, the intensities, and the carefully curated colors of red, green, black, and white. Mixed, they create a curious chromatic cacophony that remains with me. And the smell of blood brings me back to that day.

Being inducted as a Bahraini through 'Ashura' was not something I expected. But afterward, I felt like I could conduct my research with greater ease. It was as if I had passed my qualifying exam as a Bahraini. But what is Bahraini to one Bahraini is not Bahraini to another. Navigating the multiplicity of Bahraini identities was not easy. Alireza, one of my interlocutors, rested his hand on my arm and whispered, "Soon you'll have to choose." Alireza explained that every Bahraini has a position, either earned or inherited from their families, and that I was nearing the point where I would have to take sides.

Even once I felt more at ease in Bahrain, fieldwork remained complicated. Data are rarely available from a single source and it often requires patience and persistence to find out multiple perspectives. On one occasion, for instance, I learned of a government document that might help with my research, but I needed to have the permission of the author to see it. After many inquiries, I located its author, who was long retired. I visited him at his home and when I told him the purpose of my visit—to obtain a copy of the document—he retorted, "I wouldn't know where you would get that information." But he invited me to sit and "have some tea." When I was leaving, he called after me, "Come back in a couple of weeks, and we'll have some more tea." I returned several times and enjoyed my visits and never was the document mentioned again between us. On the fourth visit, however, there was an envelope on his desk with my name on it. It was never handed to me directly but when I was leaving my interlocutor's eyes rested on the envelope. I picked up the envelope and took it with me. I opened it as soon as I left his company, and inside was a photocopy of the document I had been searching for.

Having endless teas and coffees in ministries and coffeeshops was an essential part of my landscape fieldwork. It speaks to the humanity of fieldwork. Many of the interactions I had with my interlocutors were based on friendships. On trust. And trust that is built up over time. Rarely would an interlocutor open up to me on a first visit. It took time to get behind the official layer, the public narrative, to move from Erving Goffman terms the "front stage" to the "backstage." Front stage is what we present to the world, and the backstage is "a place relative to a given performance, where the impression fostered by the performance is knowingly contradicted as a matter of course."[36] I genuinely enjoyed my meetings with my interlocutors and getting to know them, and my interactions with them transcended any desire for obtaining data. In hindsight, the foregrounding of the social relations over obtaining data was important. Yet, I would likely not have been able to obtain the data in any other way.

Writing Up and Fact-Checking

Once all my research was completed, it was time to write my ethnography. Luckily, I had already started the process while in the field. Writing begins from the beginning of a research project, as the Milan-based sociologists Giampietro Gobo and Andrea Molle point out.[37] Indeed, the landscape architect Anne Whiston Spirn sometimes writes fieldnotes while immersed in the field site, giving them an immediacy and richness that would be impossible to recreate if written elsewhere.[38] Ethnography is both the act of gathering data, and the writing up of that data. I wrote most of my copious fieldnotes at the kitchen table in the evenings, in notebooks I obtained from a shop in the Manama suq under the glare of a lamp purchased from Ikea Dubai. My fieldnotes were written from the scratch notes I had maintained during the day in pocket-sized notebooks, as well as from analyzing my sketches and photographs. It worked best for me to write my fieldnotes by hand because of the immediacy of handwriting. Others find it easier to type. A discipline I learned in the field was to read my previous notes regularly, ideally every day. Although it could be tedious, I found that in doing so I recognized connections I had not seen before.[39]

I was armed with my notebooks, many photographs, sketches, doodles, and memories, but how to begin to organize them into a singular narrative? Back in Cambridge, Massachusetts, I felt removed from the world I had fallen in love with. I missed Bahrain, and my life there. I longed to walk again through the islands, smelling the salty humid air, seeing my friends, and meeting new ones. Fortunately, the deep concentration that writing takes would transport me back to Bahrain. So, I developed a system: I would write first thing in the morning till lunchtime. The rest of the day would be used for other tasks, including reading. Learning from a practice developed during my fieldwork, the last thing I would do at night was reread the text I was working on. I'd let my subconscious do the hard work overnight, then rise in the morning and begin writing again. I found the zone between sleeping and waking to be an incredibly productive space.[40]

Fortunately, I was also able to make several return visits to Bahrain to check the facts for the book. These visits were important not just to ensure the accuracy of my data, but for "self-calibration." I needed to ensure that I was not deviating too much from reality being cooped up in Cambridge. Checking data and facts required meeting with my interlocutors and renewing friendships, but I found Bahrain quite different when I returned. Bahrain was profoundly affected by the series of uprisings that comprised the Arab Spring in the early 2010s. Antigovernment demonstrations in Bahrain followed protests in Tunisia, Egypt, Syria and Yemen, persisting until 2014. There are few sizeable public spaces in the city-state, so demonstrations centered on a central roundabout called Pearl Roundabout, close to my former lodgings. The Pearl Monument in the center of the roundabout was demolished in 2011, as the government quashed the rebellion. But for years afterward, the Pearl Roundabout was blocked off, generating much gridlock in the tiny city-state. In the press, the roundabout was described as the Pearl Square. From the perspective of international media, it would seem revolutions don't happen on roundabouts. Bahrain has since reconfigured its traffic arrangements to use junctions, which are easier for authorities to control.

When I returned, many people remarked how I would not be able to do the fieldwork that I had done before. On my first visit after the Arab Spring, I was approached by four men wearing black t-shirts when I took out my camera in the suq. They asked me to put my camera away "for my protection." My interlocutors explained that walking around the island would also now be much more difficult for me because of the increased police presence and suspicion of foreigners. Others refused to take me to the more rural areas of the islands, or "villages" as they are locally called, the locii of Shiʿa uprisings in the past. These events taught me that broader historical, political, and economic circumstances and contingencies outside of the researcher can affect fieldwork, making certain kinds of research possible or impossible.

Indexing

As I set out with the seemingly immense task of giving form to my fieldnotes, I looked for a structure for the ethnographic text I planned to write. My notes were arranged chronologically, but if the ethnographic text were chronological too that would simply reproduce what I had observed and interpreted in the field. I needed another structure based on the interpretation of my notes. As I reread my fieldnotes, I found three main themes that emerged: politics, economics, and social life as they pertained to the green landscape. I read my fieldnotes again, this time noting moments and paragraphs that I felt had some importance for the eventual narrative. By hand, I transcribed relevant vignettes to index cards, using different colors of cards to represent the three overarching themes of politics, economics, social life. The indexing process made those vignettes stronger in my memory. I then lay the cards on the floor of my apartment and began to trace associations among them and organize them into three chapters on the politics of green, the economics of green, and the social life of green. This exercise quickly demonstrated how central politics was in terms of greenery in Bahrain; that category had the greatest number of cards, accounting for over half of them. Clearly, this structure would not work. Either the political was going to be half the book, or it needed to be embedded within the other categories.

So, I started to look for another structure—a taxonomy. The architectural ethnographer Galen Cranz describes taxonomy as "the organization of categories" and as "a logical structure of meaning."[41] Indexing in order to create a taxonomy was also a process of editing, of deciding what was important to include in the narrative and what not to include. At one point after I had written the ethnography, I estimated that 90 percent of my fieldnotes did not actually make it into the book.

As I struggled to find the right taxonomy, I found myself doing additional reading on color. It was already clear to me from my research that one cannot understand a color without knowing how it relates to other colors too. This important principle of color theory is illustrated by Josef Albers in his foundational book *Interac-*

tion of Color. Albers writes: "In visual perception a color is almost never seen as it really is—as it physically is. This fact makes color the most relative medium in art."[42] Albers designed a series of exercises to demonstrate the relationality of colors that are documented in his book. I had been familiar with Albers's arguments, but while I was writing I returned to his work in visits to the Faber Birren Collection of Books on Color at Yale University and the Colour Reference Library (CRL) at the Royal College of Art in London. This was another turning point: I realized if color was how Bahrainis understood and conceptualized landscape, surely color could provide the structure for the narrative?

It was clear from my fieldwork that Bahrainis conceptualized and understood green in relation to the desert around it. Green takes part of its meaning from the beige desert, not to mention the blue sky, beige buildings, and turquoise sea. In addition, my research had shown that red was not only a complementary color to green but was often used as a metaphor for green, or what I term a chromatic metaphor. A chromatic metaphor is when one color, in this case red, is substituted for another color, in this case green, adopting many of green's uses and meanings. In Bahrain, green is often associated with resistance to the state, so the state promotes red instead of green.[43]

So, I set about indexing my fieldnotes according to colors: blue, red, beige and, of course, green. The big question was: Which green? I saw two primary greens: the olive hues of the date-palm green, versus the bright greens of lawns and roadside strips. I lay the revised index cards on my apartment floor and traced associations. In arranging the index cards according to color, the chromatic geographies became more visible. This practice confirms an insight from James Corner: "Taxonomic and genealogical procedures of relating, indexing, and naming can often be extremely productive in revealing latent structures. Such techniques may produce insights that have both utility and metaphoricity."[44] Indeed, arranging by color became the eventual structure, because I realized one cannot understand green without understanding these chromatic relationships. This method determined the arrangement of the book, with each chapter focused on a different color, beginning with blue and then date-palm green, grass green, red, beige, and white, as the coming together of all colors. This experience led to further experimentation with indexing in later projects.

Sharing the Research

Gobo and Molle draw attention to one of the difficulties with ethnographic writing being that you write for two audiences. One audience is the academic community who will read your work, and the other audience is made up of your interlocutors, who can be your biggest critics.[45] As part of my writing-up process, I had several Bahraini residents and scholars read my work. I was careful to have Bahrainis from different social groups read it. They were indeed great critics, invested in ensuring the accuracy of my descriptions. My friends and interlocutors generously helped me check and, when necessary, revise facts, phrases, and idioms.

And then there was Aly. Aly was a senior figure in one of the ministries, and I gave him my draft text to read. When we met to discuss it, he demanded, "You cannot publish that book!" He told me he did not even want to be in the same room as my text. Aly said I was naïve and accused me of quoting what random people told me in the suq. I explained that I had consulted with many people, and carefully reflected on what they had told me. But he said that's not how it comes across. It seems the main problem for him was that he felt I was insulting the king with some negative comments I repeated about the ministries and the Economic Development Board, even though all the information was already in the newspapers. Aly seemed to think I was critical of the Bahraini state and, knowing that I am Irish, Aly claimed that Ireland was one of the most corrupt countries in Europe. "Compared to Ireland, Bahrain is a paradise!" he roared. Aly said I was giving fuel to people who want to destabilize the state and they would use this to advance their faulty arguments. He said it was the single most political document he had read in recent years, and that he would defend Bahrain "to the death." His fierce critique went on for over three hours. He warned me if I published the book, I would not be allowed back into Bahrain; I would be forever banished. When I left my meeting with Aly that day and left for the airport, I sensed we would never meet again. I wasn't even sure if I would meet Bahrain again.

Aly's intervention was counterproductive to his official narrative. Because he threatened me, not because he was necessarily arrogant but perhaps more so because he was frightened, I could not change my manuscript to suit him. To do so would have been an affront to academic freedom. Indeed, after our meeting I included more criticisms of the state, and I also included positive statements too. And, later, when I met members of the royal family, I did not receive any negative critique—on the contrary, they were positive about the book. When I recounted my meeting with Aly to a member of the family, I was told, "Don't listen to him. *We* like what you do!" I thought this quote was interesting because it suggests *they* know what I do.

There were many aspects to the publication process that there is not time to go into here. But I do want to speak about the images. While I took many photographs, I also needed maps, diagrams, and historic photographs to reinforce the points I wanted to make. One image eluded me, a map that apparently referred to the Indian Ocean as the Green Sea. I had been told about this map by a professor at the Gulf Studies Conference I attended at the University of Exeter, but he was unable to direct me to the source, and neither could I find it. So, I enlisted the help of a research assistant, Sunny. Over the course of about three months, Sunny visited the Harvard Map Collection once or twice a week, sifting through the archives, but found no sign of the elusive map. Eventually, I asked Sunny to stop, as my research budget was running low. A few weeks after that, Sunny messaged me to say she thought she had found something. Sunny was in Harvard's Houghton Library and had found an even more relevant map, titled "Map of the World after the Destruction of the Tower of Babel." The map terms the Persian Gulf, the *Sinus Persicus* (Persian Gulf) and *Arab Mare* (Arab Sea), and also as *Alachdar* (the Green), stating "*id est viride*" (it is green). Even after I told her to give up, Sunny kept looking for the map on her own time. Landscape fieldwork demands relentlessness.

Strengths and Limitations of Ethnographic Fieldwork

To return to the question raised at the start of this chapter: Was long-term ethnographic fieldwork worth the time and money that it took? My answer is yes, but with a few caveats. I'll start with the main strengths of the method. Long-term fieldwork allows us a way to create new knowledge, including discoveries that were not on our original agenda. Through "embodied engagement,"[46] ethnographers immerse themselves in the everyday and the ubiquitous, which might seem extraordinary and strange to them.[47] We learn the local language, dialect, modes of communication,

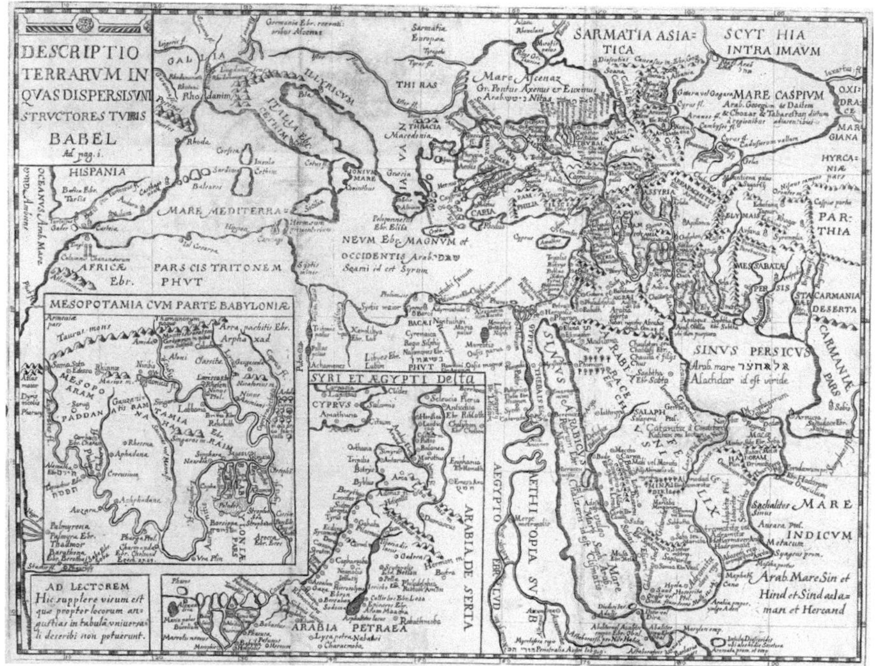

Figure 50. "Map of the World after the Destruction of the Tower of Babel," showing *Sinvs Persicvs, Arab Mare, Alachdar, id est viride* ("that is green"). Created in 1681, printed for Johannis Davidis Zunneri after the death of Samuel Bochart (1599–1667), cartographer. (*Geographiae Sacra*, 1681. AH 3716.5, Houghton Library, Harvard University)

and social norms. My life in Bahrain involved endless meetings, meals, and coffees with my interlocutors, as I learned how to be Bahraini. At first, this process was the exact opposite of my life in Ireland, but the longer I stayed in Bahrain, the more familiar that place too began to feel. When not having meals with others, I would return to my own favorite cafes and restaurants, such as Jahan Grills in Gudaibiya, and the former Nirvana in the suq, to achieve a sense of the familiar and to know a few places very well rather than a lot of places superficially. In all my subsequent projects, I have had equivalents of the Jahan Grills.

Long-term fieldwork allows the ethnographer to learn how people dwell in their landscape by inhabiting it themselves. The main point is to be present for "non-events," the patterns of the everyday, although the extraordinary, of course, also interests us. But all of this takes a tremendous amount of time that not all researchers or practitioners have. The rule of thumb, as I noted earlier, is that when writing fieldnotes we spend four hours interpreting every hour in the field.[48] Not only is writing notes a laborious process but rereading previous fieldnotes repeatedly to look for connections, relationships, and meaning also takes time. Meanwhile, we need to bear in mind that our interpretations are rarely neutral. They are informed by the life experience of the ethnographer as much as they are by the fieldwork, so interpreting one's own observations requires considerable time to properly fact-check and reflect on what it all means.[49]

Long-term fieldwork can also be mentally and emotionally trying. Despite the constant contact with people, ethnography can also be a lonely and tedious process. The ethnographer spends hours, days, and weeks in the field when seemingly little happens. As I found in my early days in Bahrain, it can be hard to keep fieldnotes when it seems you are not making progress. But at those times of seeming stagnation, the subconscious continues to work, interpreting the minutiae of everyday occurrences and the bigger picture. One must be patient. Connections sometimes reveal themselves in strange ways and at the most unexpected times. It's also hard to write notes when so much is happening: at those moments, one must slow down and find time to write.

The relationships we form with interlocutors during fieldwork are often the sources of the most important insights we make, but these relationships can also be difficult to navigate at times. Our interlocutors train us to be participant observers in their lives. At the same time, our interlocutors can become our friends as we realize our shared interests and passions. This leads to a dynamic where one must make judgments about what is personal and what is done in the name of research. For example, my conversation with Isa that evening in the opening of the chapter was not about my research. It was a conversation between friends chatting in a car. Because we were talking as friends rather than interviewer and interviewee, I did not write about the conversation in my fieldnotes—although it remains to this day in my headnotes, where it will stay. Other ethnographers might treat the encounter differently by keeping a journal alongside their fieldnotes and meticulously documenting both the formal research happenings of the day but also the reactions and feelings toward them.[50] The fieldnotes would be the business side of things and the journal would record personal conversations and considerations. In the end, I drew upon it all: emails and fieldnotes, notes and photographs, scratchnotes and headnotes, sources and endnotes. All the while, I tried to keep my own prejudices in mind, avoiding orientalism and my predilection to fetishize my life in Bahrain.

Further Research Questions

The luxury presented by a year walking through Bahrain's landscapes, and time afterward to think about it, allowed me to propose two possibilities for landscape architecture in *Paradoxes of Green.* First, I argued that designers should address color more centrally in their work, given the paucity of literature on the topic. Second, I called for designers to better engage with our audiences, and to consider how to write for them. How do designers understand human ecologies, never mind chromatic ecologies? One is at the level of color, and the second is through the level of community. But even years later, my year of fieldwork in Bahrain continues to throw up new ideas and research leads. Two questions in particular stand out. First, what is the agency of color to shape the landscape, especially in cities? Second, if green is not always suitable for the environment in arid regions—due to its need for irrigation and maintenance—how can we use green more strategically in beige landscapes like Bahrain's?

The answer is not to not use green. Time and time again in Bahrain, my interlocutors told me of the importance of green, politically, economically, culturally, and socially. I was told how every year the ministries desperately struggle to stop green grass turning brown with the summer heat. In the process, they expend vast amounts of financial and human resources to keep the green space green. When ministries try not to provide green space, along roadsides and in new residential developments, they are inundated with complaints from the public saying how ugly the desert is and how they need to have the green. When the ministry attempted to replace water-guzzling grass and shrubs with more desert-loving species, which generally happen to be brown and beige, this was overruled by the those in positions of authority who said, "We have no mandate to promote that image."[51] For Bahrainis, green is as necessary an element of the city as are sidewalks and curbs. This necessity is derived from the relational element to beige, and the promise of a greener world. This verdant aspiration is no doubt driven by the development of fantasy related to images of what constitutes a global or contemporary city. Singapore is often cited as a verdant model for Bahrain, but it has a very different climate, one that can sustain the greenery.

That returns us to the question of how to then use green more strategically and in a more environmentally responsible way in cities. My fieldwork in Bahrain taught me that first you must know which green you are referring to. The gray-greens of the date palm groves used to be the greens of the elites, but they have been replaced by the bright greens of new roadside developments and golf courses. The values of green have been changing. In addition, we cannot understand green in isolation: green exists in relation to other colors and hues. If we consider that green's agency is informed by the colors around it, then the answer to the green conundrum is not with green but with other colors. The full spectrum of the rainbow offers greater hope for environmentalism than the German greens of the 1970s.

Inspired by the chromatic studies of the artist and Yale professor Josef Albers, in *Interaction of Colors,* can we use other colors to amplify the impacts of green? For example, by painting a wall white adjacent to a green area, can we reduce the area of green adjacent to the wall without overly affecting green's overall impact? Or, by painting a wall green adjacent to the desert, can we compensate for the soft greens of vegetation and avoid greening the desert? Using collage to analyze arid urban landscapes and propose alternatives that minimize green and maximize the chromatic impact, we can ask if it is possible to envision preferred future scenarios for a less green, more colorful landscapes.

Meanwhile, a recent study from Yale University illustrates the potential of color as a direct response to issues over climate change. The study suggests that white roofs can deflect heat and keep cities cooler.[52] The same study tells us that dark surfaces absorb more heat from the sun: "Fresh asphalt reflects only 4 percent of sunlight compared to as much as 25 percent for natural grassland and up to 90 percent for a white surface such as fresh snow."[53] To this end, the city of Los Angeles has been painting dark asphalt road surfaces with a light gray color. This has the impact of reducing temperatures by up to 10 degrees.

One takeaway from my project is that landscape architects need to work within the codes and behavior of their interlocutors. If they do not, there is an epistemological inconsistency. Landscape architects designing in Bahrain, most of whom are foreigners, often have one idea of landscape in mind, whereas Bahrainis have another. And then there are indigenous landscape practices, such as farming among the date palm groves and the complex irrigation systems. I am often embarrassed by much of what is constructed in Bahrain in the name of landscape architecture. Time and time again we see green grass parks with water-guzzling vegetation designed to look good in photographs or from passing cars. Rarely do they have human practice in mind or consider how people will relate and use the spaces.

The Death of the Groves

After my year of fieldwork, I returned to Bahrain on several occasions to check my data as I wrote up the ethnography. On one of these visits, toward the end of my project, I met an engineer who had previously been very helpful to me. Our interaction is described in my book *Paradoxes of Green: Landscapes of a City-State.* The engineer told me how the treated sewage effluent project they had initiated had resulted in the destruction rather than the saving of the groves.

> So it was that Bahrain's remaining date palm groves, continuous over centuries, were deemed to have been lost in a few short years.
>
> The solution to keep the groves alive . . . was a purely technical one, perhaps even a green one, but it did not account for human practice. Had the engineers spent time in the groves, learning the codes and patterns of behavior of the farmers and of the trees themselves, they might have become aware of the history, dreams, and ambitions within those groves. They might have realized that it is difficult to resolve an ecological problem with a solely technical approach. Green cannot function long term as an engineering solution without accounting for its economic, environmental, political, and other symbolic meanings. Had the engineers taken a "greener" approach, the date palms might not have been lost.[54]

The changes seriously challenged my belief in ethnography. I was devastated to learn of the loss of the date palm groves—spaces that I had come to love profoundly—and at the collapse of the agricultural industry. Deep down I was not surprised, though. My fieldwork had demonstrated that agriculture in Bahrain was practiced for aesthetic and emotional reasons, and not for profit.[55] It is much more economical for fruit and vegetables to be imported from Saudi Arabia and other countries than to grow them in Bahrain.[56] And, clearly, the groves were being threatened by developers, who wanted to replace the existing date palms with villas, and who claimed they would replant the date palms. I had also become familiar with the narrative from many Shiʿa that the Sunni-controlled state had no interest in saving the groves, as they allegedly saw them as sites of Shiʿi resistance. While I did not get a sense from the engineer that treated sewage effluent was a deliberate device to destroy the groves, it is not a far-fetched assumption. So, while sad, it was not unexpected that the groves had been depleted to such an extent. The speed and extent of their destruction was surprising to me, though.

The experience left me with a nagging question: Did I take too long? Could I have intervened to save the groves somehow if I had moved faster? I had followed the classic ethnographic process: a year in the field, a couple of years writing about it for

Figure 51. Death of the groves, Sanaḍ, Bahrain, 2022. (Photograph by Camille Zakharia)

my doctoral dissertation. The overall process took ten years from initial fieldwork to printed book. And this is not unusual. But as I was thinking, writing, and dreaming about the groves' significance, they were being lost.

However, this questioning was unfair in many ways and downplays the role of the ethnographic process. The myriad hours I spent discussing the groves in the ministries and with Bahrainis must have had some resonance. For example, at one point I was asked to prepare a PowerPoint for a meeting with the crown prince, and then given the wrong time for the meeting. The person who gave me the wrong time then delivered the presentation I had prepared. Saving the groves was not a major goal when I set about doing my fieldwork, but the pressures on the groves emerged from my fieldwork as one of the most urgent issues facing Bahrain's landscape. I tried to think of mechanisms for their conservation. At one point, I had recommended they become a national park.

Reflections

I have described a year of ethnographic fieldwork in Bahrain, some challenges and opportunities it presented, and questions it opened. I introduced four stages of my fieldwork in Bahrain: remote and preliminary fieldwork, acclimatization, becoming, and writing up and fact-checking. These four stages can be applied to other fieldwork projects, insofar as fieldwork is not just limited to the time in the field. I show the value of a single word ("green") and how through seemingly minute details, we can come to understandings that have more extensive spatial reverberations. I introduced the power of landscape fieldwork to challenge norms and preconceived notions on the meaning of landscape and demonstrate that landscape is not a universal concept. To work in a society, one must know the nuances of meanings. Landscape fieldwork gives us a way to understand those nuances and to continue to question them.

In the year I spent walking through Bahrain, I came to know its landscape intimately and love it. I took several years to check facts and write my book. Meanwhile the date palm groves I loved so much were lost. For me, this created an enormous challenge. While ethnographic fieldwork has its limitations, it also reminds us of the value of design, as a descriptive/prescriptive, projective/proactive endeavor, and its potential to address urgent environmental issues. Design and planning are based on different temporalities than academic research. So, it seemed to me, the only way to overcome this issue of temporality was to share the fieldwork among several fieldworkers rather than just one. To this end, after my experience in Bahrain, I began to ask myself how processes of ethnographic fieldwork could be removed from the single author and shared among a collective. I realized such a method could also address persistent questions over perceived biases embedded in ethnographic research. This collective methodological question is addressed in chapter 4. I found that it's necessary to balance the benefits of long-term ethnographic fieldwork—the ability to make deep insights and form relationships with community—with the more fast-paced and proactive nature of design that allows us to imagine preferred futures.

Still, I felt I was not able to do enough with ethnography. My frustration with the harsh reality of the groves' demise inspired me to imagine ways to do landscape fieldwork more quickly that might more directly benefit design and planning processes. Other ethnographers are currently grappling with the same issue and attempting to invent new means of doing ethnographic fieldwork that could respond more quickly to urgent problems.[57] Others are working on making research more accessible to the public—as evidenced through the growth of multimodal approaches such as film and graphic ethnographies, open-access efforts, and publishing on blogs and other social media.

4

ENGAGING COLLECTIVELY

Collaborative Fieldwork in The Bahamas

You ain't hurryin' me, my brother, you ain't hurryin' me.
I don't want no heart attack, or ulcer in my belly.
—Eric Minns, "A Bahamian Calypso"

"Stop! Stop! Stop! I'm running a country, and you're talking to me about chicken coops? What can *you* tell *me* that can help *me* make better decisions?," the prime minister roared, pounding his fist on the table. I was just starting to present the results of a research project to the cabinet of The Bahamas. Sensing my unease, the deputy prime minister began to explain the value of the chicken coops we had built on three islands. The prime minister leaned toward the deputy prime minister and said, "Shut up!" Then, pointing right at me, he yelled, "I want *him* to tell *me!*"

Two colleagues and I were meeting with the cabinet that day to formally present the results of a sponsored research project conducted in The Bahamas. The Harvard Graduate School of Design (GSD), where I teach, had been asked to work with the Government of the Commonwealth of The Bahamas and the Bahamas National Trust (BNT), who advise the government on issues over sustainability and climate change. Our task was to imagine the sustainable development of Exuma, an archipelago of 365 islands and cays southeast of New Providence, the island where the capital of The Bahamas, Nassau, is located. The project was based on a book I coedited with the then dean of Harvard Graduate School of Design, Mohsen Mostafavi, called *Ecological Urbanism,*[1] which was inspired by Félix Guattari's *The Three Ecologies.*[2] The project proposed integrating environmental, social, and aesthetic ecologies, rather than treating these fields separately, as a fundamental aspect of sustainable design and development. The project in this district of The Bahamas was formally titled "A Sustainable Future for Exuma: Environmental Management, Design, and Planning," or "Sustainable Exuma" in short. As the co–principal investigator, I had the task of opening the presentation of the multiyear collaboration to the government of The Bahamas.[3]

As calmly as I could, I replied that on an island called Little Farmer's Cay residents told us they had forgotten how to farm. Over the previous fifty years, the soil had blown away in hurricanes. In fact, the tiny island of sixty residents didn't produce much anymore, relying instead on supplies coming from the outside. Every week, the mail boat would arrive to the island laden with mail, food, goods, and other necessities, then return to New Providence empty. Residents had also told our fieldworkers that the island was full of chickens; they were woken in the early morning by the roosters' crowing. But the chickens were wild, and no one could find their eggs.

I further explained to the prime minister that on another island the school principal maintained that the curriculum prepares students for a life off the islands. The teacher explained that the primary school curriculum favored the 70 percent of the population that live in the capital city of Nassau, where we were meeting that day. But the curriculum did little to prepare children for a life on the islands and cays. "They've no idea where food comes from," he told us. I reminded the prime minister that at any moment in time, The Bahamas has two weeks' supply of food and that most of that food comes through the United States. After the attacks of 9/11, the United States closed its borders and The Bahamas nearly ran out of food. And, of even greater long-term significance, I informed the prime minister that islanders seemed oblivious to rising sea levels and extreme weather events—the same sequence of events that blew the soil away from Little Farmer's Cay in the first place. What's more, much of the land on these islands is expected to be submerged in the

Figure 52. Chicken coop. (Sustainable Exuma)

next thirty years on account of rising sea levels, with the sea both inundating the coastline and percolating up through the limestone bedrock.[4]

The chicken coops, which had been constructed a year before our meeting, were one of our project's responses to these various challenges we had learned from our fieldwork. I reminded the prime minister that these pressing national issues over agriculture, education, food security, and climate change were all matters that were directly under his control. The chicken coops that we built as a very small landscape fieldwork project to impact the everyday lives of islanders incorporated aspects of decisions made at the national level around that very same cabinet table.

With a gentle nod of the head, the prime minister acknowledged my answer and his and the government's agency in all these issues. I then proceeded with the formal presentation. In the discussion that followed, our conversation ranged over many topics, from climate change, to transportation, to governance, which all came back to the political agency of landscape architecture. Too often landscape architecture is left out of political decisions, yet landscape by its very nature is political.[5] Indeed, one of our conclusions from this project was the centrality of landscape to the political life on the islands. In the end, our meeting went over time by one hour: the prime minister, after all, wanted to know how our research could help him make better decisions at the national level. When we left the cabinet office that day, there was a line of people waiting to meet with them who had been delayed by our discussion. I felt sorry to have kept everyone waiting, but relieved and excited that after such a stressful start the meeting had gone well. We'd been able to spend two hours discussing the role of government in landscape curation, and the role of landscape architecture in governing—a sign that our project, based on collective landscape fieldwork, was accomplishing what we set out to do.

To answer the questions, we had developed a new form of landscape fieldwork based on collective engagement. This project first came about at a time when I was teaching a course titled "Design Anthropology: Objects, Landscapes, Cities," cross-listed between the Departments of Landscape Architecture and Anthropology at Harvard University. The course was predicated on the notion that the best way to teach students how to do fieldwork is for them to do it. The project allowed us to bring students to The Bahamas to do fieldwork *in situ.* While this may seem a great luxury, and it was, the fieldwork component was a small percentage of the entire budget for the project, the rest going for salaries, and expenses associated with executive education and workshops. I think most people would agree that the fieldwork was a major percentage of the project's impact. I mention this because fieldwork is often dismissed as beyond the scope of most projects, professional or otherwise, but

Figure 53. Building a chicken coop. (Sustainable Exuma)

our experience suggests otherwise. This chapter outlines the method that we developed to move from a form of landscape fieldwork based on *description* to landscape fieldwork as *prescription.*

A Centralized Landscape

The prime minister's point about why I was talking to him about chicken coops is not unreasonable or all that surprising. I was talking to the prime minister about the chicken coops because there was no one else to talk to about them. The Bahamas received its independence from the United Kingdom in 1973 and, to this day, has a highly centralized governmental structure that includes a parliamentary system, civil service, and bureaucracy inherited from the British. A senior official explained that the state was Nassau-centric because it was from there that the British exported the various goods, such as sugar, timber, and rum: they extracted resources from the smaller islands, then took them to Nassau, then shipped them to Southampton or London. Most of the population, as noted, live on the island of New Providence, meaning that the political class of The Bahamas is centralized too. The governor-general represents the British monarch, and power is centralized in one seemingly all-powerful prime minister.

The system of governance in The Bahamas means the national government is responsible even for small chicken coops, as much as they are for school curriculum and planning policy. Indeed, even though the coops might seem insignificant, they embody a complex assemblage of relationships between operational fields, institutional partners, community groups, and the spaces of the islands. We were charged with developing innovative ways to complement professional practice and the work of government under the framework of imagining alternative futures for the Exuma archipelago. Before the project began, we had several rounds of negotiations between the various parties, including two conferences that allowed us to think through the scope and ambition of the project.[6]

The site of our focus was, as noted, the archipelago of Exuma, which, according to our estimates, has 365 islands and cays. (It can be difficult to classify what an island is, given that some are just rocks in the sea, but we counted 365 and tried to learn about one of these islands per day over the course of a year.) The Bahamas comprises

Figure 54. The 365 Islands and Cays of Exuma with the Exuma Cays Land and Sea Park highlighted. New Providence is in the top left. (Maria Vollas)

700 islands over all, and is divided into 32 administrative districts, with 110 councilors and 281 town committee members. Overall, there are 55 parliamentarians, with just one member shared between all of Exuma and Ragged Island.[7] In total, there are only 7,000 residents in Exuma, or about 2 percent of the total population of The Bahamas. As we learned from our fieldwork, although political power officially flows through the office of the prime minister, the government is in some ways, paradoxically, not all that powerful in Exuma. Many people told us how Exuma was so far removed from the workings of government that anyone who had any money could effectively do what they wanted.

Most of the Exuma islands are uninhabited, and many of the islands don't, at least yet, have names. The islands have distinct ecologies: certain species of plants and animals live only on specific islands, and the same can be said for people. But one thing they have in common are the depleting soils, a phenomenon that has been ongoing for fifty years or more, according to what people told us.

In the Exuma archipelago, which extends over 120 miles (approximately 190 kilometers), we focused on the main settlements of George Town, Rolleville, Staniel Cay, Black Point, and Little Farmer's Cay. We also spent time on Cat Island, one of the sites on which Christopher Columbus allegedly first landed in the New World in 1492,[8] and visited Long Island nearby to the south on account of its land tenure system and the robust local responses to hurricanes. George Town (population ca. three thousand) is the capital of Exuma and fourth-largest settlement in

Figure 55. Islands and Cays of Exuma. (Marissa Wells)

The Bahamas. Named after the English King George III, George Town is located on Great Exuma, the archipelago's main island to the south, and is home to about half the population of the chain. George Town is midway between Rolleville to the north, and Williamstown to the south, and home to the main airport, Exuma International Airport. To the north, there are the settlements of Staniel Cay, Black Point, and Little Farmer's Cay with populations of approximately two hundred, one hundred, and sixty respectively. They are interspersed with private islands of the mega-rich. As the beating heart of the Exuma Cays, Staniel Cay has the busiest landing strip in Exuma and is home to a variety of tourist attractions. These include Thunderball Grotto, which gave its name to a 1965 James Bond movie, a group of fifty to sixty swimming feral pigs, and the Staniel Cay Yacht Club, the social center for boaters in Exuma. A person could spend a whole year in Exuma, and still find it challenging to learn about all these different islands and island communities. Fortunately, we had multiple researchers working over a period of three years on the project. In the end, more than fifty-two individuals each engaged in at least one week of fieldwork in Exuma. Each fieldworker was trained in ethnographic techniques before going to the field, and the individual's fieldwork became part of a collective experiment, which asked this question: Can fifty-two fieldworkers, each spending one week on site, equal more than one fieldworker spending a year?

Welcome to Paradise!

Arriving in the island of Staniel Cay by propeller plane, one is invariably greeted by the phrase, "Welcome to Paradise." The local government representative generously makes a point of greeting every arrival on the island in his golf buggy. I was struck by how many times I was welcomed to Paradise during the four years of the project.

The entire archipelago is a paradisiacal-looking landscape, for sure. Astronauts report that The Bahamas stands out from space: they call it the "jewel of the earth" because of the distinctive arrays of blues and greens in the waters. The sea surrounding Exuma is indescribably beautiful, with all manner of shades of turquoise and sapphire that dazzle in the Caribbean light and interact and mingle with the sky. Exuma sits on a narrow shelf, and on the western side these fantastical hues are often attributed to the shallow seabed, usually just a few feet deep; in Spanish, *baja mar* means “shallow sea,” which led to the name, The Bahamas. These colors are also affected by the light color of the seafloor as well as the light conditions.[9] Meanwhile, the seabed on the Atlantic side, to the east of Exuma, is deeper, and consequently the waters are darker blue and rougher because of the ocean. Boats generally prefer the Atlantic side as it is easier to navigate. Buildings in Exuma are often painted turquoise, or a shade of blue, to reflect or contrast with the sea’s colors. Sometimes they are painted pink, invariably with the particular shade of pink found on the inside of conch shells. Conch salad is the quintessential Exumian or Bahamian dish. These chromatic traditions integrate the colors of the landscape within the architecture, which is often deferential to the landscape. In Exuma, the landscape takes all the attention.

What Is It That You Do?

During the project, I introduced an eminent European landscape architect, whose name I will not mention, to the landscapes of Exuma. He exclaimed that Exuma’s landscapes are so beautiful that it was impossible to imagine what a landscape architect can possibly do there from a design perspective. I was surprised at his feelings of inadequacy in the context of the extreme beauty of the islands. I was also somewhat flummoxed at the implication that landscape architecture is about beautification. Indeed, it is true that human interventions on the islands generally disrupt the harmony of Paradise. In this sense, the eminent landscape architect is right to wonder about the efficacy of the profession to compete for attention with such an overwhelming landscape. It is telling that there are few landscape architects in The Bahamas and none in Exuma. Noted landscape architects, such as Raymond Jungles, have worked in Exuma but on private islands. Public landscapes are untouched by landscape architects. If a landscape architect were to design in this sort of environment, it would be, in many respects, the inversion of what most landscape architects do. Often landscape architects work on a landscape and make it more beautiful. For example, landscape architecture is often seen as a remediation of the impact of urbanization. Perhaps in Exuma the landscape architect needs to focus more on functionality, on producing food, on gardening and remediating the impacts of an unproductive, albeit stunningly beautiful, landscape.

As I got to know the islands better, however, I became a little bit less convinced that where I was being welcomed was Paradise. The profoundly heartfelt greeting became troubling. As one visitor to the islands remarked, "There's only so much looking you can do at a beautiful beach." The same individual explained that the beaches and the heat become oppressive, forcing people to stay indoors. They complained of stores selling little more than rice and hard liquor. Exuma is not always so beautiful for the insiders, either. There are, in some respects, two extremes of people who live in Exuma: those who grew up in Exuma and have few other options open to them, and the superwealthy who can afford the additional costs of maintaining a luxurious lifestyle on the rustic, seemingly remote, islands. As the British social anthropologist Edin Ardener brings to our attention, what we consider remote areas are not always as remote as we think.[10] The island lifestyle also attracts a third group of people, those who can afford a second home and live in Exuma until something goes wrong, or illness or incapacitation drive them away. And then there are the tourists, who form the main industry on the islands, and the low-income immigrants who quietly, and largely invisibly, help with the tourist and construction industries.

Among these groups, there are substantial social inequities; there's not just a lack of cash, but a lack of steady incomes that would generate the cash. Alcoholism is not uncommon on the islands. Island residents live without many of the modern services that others take for granted in contemporary life. The island of Staniel Cay receives electricity through a single cable that runs along the seabed from the nearby island of Black Point. Without a steady electricity supply, credit card and cash machines don't always work, making everyday commerce challenging. Higher-level education mostly takes place off-island, on New Providence. The school system requires teenagers to leave for Nassau, where they might lodge with relatives, usually never to return full-time to Exuma. Health services too are limited; even in death, Exumians are transported to the mainland for autopsies and often for burials. Admittedly, though, many see the slower pace of island life as an advantage. Our research indicated, for example, that heart disease is much lower on the islands than the national statistics—probably due to a diet rich in fresh fish and low in processed foods. But suffice to say that we realized during the project that any kind of design intervention we might make should embody a low-cost efficiency that would work within a fragile economy. So, we set out to better understand the constituencies we were engaging with and designing for.

Understanding the Constituencies

"You can't understand Exuma unless you understand the golf carts," we were told by a long-time resident of Staniel Cay. These vehicles are perfect for the narrow roads and slower pace of island life. They transport not just people but groceries, goods, and gossip. Understanding the human ecologies of the islands was as essential as understanding the species of plants and wildlife that inhabit the islands, not to mention the rocks and soils that comprise the land and the seabed.[11] In general, we lived among the small communities and tried to participate or engage in everyday life as far as was possible, both on and off the golf carts, given the many limitations that were encountered. Through our initial visits to the islands and conversations with institutional partners, we identified six broad constituencies for engagement: (1) island communities, (2) private islanders, (3) government and NGOs, (4) tourists, (5) immigrants, and (6) the environment. While on reflection, I see that these categories are inadequate, they are still helpful for making sense of who inhabits the islands.

The first, and most obvious, constituency comprises the communities that are

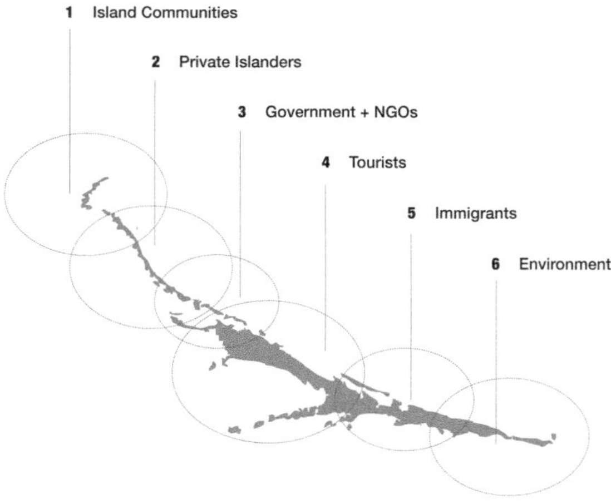

Figure 56. Constituencies for engagement. (Sustainable Exuma)

Figure 57. Staniel Cay, Exuma. (Marisa Wells)

made up of the settlements on the islands. As noted earlier, these include George Town, Williamstown, Rolleville, Staniel Cay, Black Point, Little Farmer's Cay, and so forth. These communities are home mostly to descendants of enslaved peoples brought to the islands from the southern United States by Americans and British settlers. The islands were granted by King George III to British loyalists who had fled the United States after the American War of Independence. To this day, many Exumians carry the names of British landowners, such as the Rolle family, who live on the land formerly owned by Lord John Rolle. The story goes that, having no children, Lord Rolle bequeathed his landholdings to his enslaved peoples, and the settlements of Rolleville and Rolletown bear his name as a mark of acknowledgment. Other residents include descendants of the "Eleutheran Adventurers," a group of Puritans who came to the island of Eleuthera in The Bahamas after the American Revolutionary War. Much of the land is extralegal, or outside formal legal systems. The land pattern on the islands remains a primarily commonage system. This system of ownership is shared among the descendants of enslaved peoples, and each family has a different mode of administering the land. Commonage cannot be sold. As a result, it is difficult for nondescendants to develop the land. Albeit frustrating for some, this has the effect of protecting the land from development since it is not possible, or at least complicated, to sell it off to private owners. It is one of the reasons the land is to this day largely unspoiled with contemporary developments. Exumians often described themselves to us as "land rich, and cash poor."

The second constituency includes the owners of private islands, mostly within the national Exuma Cays Land and Sea Park, described in more detail below. In the 1930s, the government started selling islands to wealthy individuals, and to this day many islands are privately owned. Exumians often expressed surprise at why anyone would spend $75 million for an island landscape that they see every day and enjoy for free.[12] Many of these islands are owned by billionaires on account of the island's perceived remoteness and The Bahamas' low taxes, and they employ staff from the nearby communities. The owners include film stars such as Johnny Depp and Nicholas Cage, and the magician David Copperfield. Copperfield's island, Musha Cay, starkly shows the economic disparities between locals and private island owners. Musha Cay and the Islands of Copperfield Bay can be rented when the host is not in residence, with rates starting at $57,000 per night,[13] far more than the average Exumian could hope to earn in a year.

His Highness the Aga Khan, who sponsored the overall Sustainable Exuma research project, has an island within the Exuma Cays Land and Sea Park.[14] These private islands of the megarich, effectively principalities, often fly their own flags and operate with their own rules. Access to these communities is restricted, highlighting some of the limitations of ethnographic fieldwork.

From our short visits, we concluded that many of the private islands embrace an aesthetic of sustainability—such as stone walls and thatched roofs—but in practice they're often heavily energy-dependent when one considers the embodied energy, including labor, that go into their production. Others have vegetation that is woefully unsuitable for the landscape, but there is no one to enforce any rules. In other cases, the islands function sustainably but are blights on the landscape. Many of our interlocutors singled out as a major eyesore the privately owned Over Yonder Cay, which was covered in wind turbines and solar panels. While the island appears self-sufficient in energy, for many other private island owners we talked with, the appearance of the island was unacceptable, and they told us it created a profound paradox in Paradise.

Third, there are the government bodies and NGOs and religious institutions who have agency over this site, in particular the Exuma Cays Land and Sea Park (ECLSP) and protected marine areas. The Exuma Cays Land and Sea Park, the first mostly water-based national park in the world, was founded in 1958 and stretches twenty miles long (thirty-two kilometers). Managed by the Bahamas National Trust (BNT), the park came about as a result of the Exuma Expedition that same year; it was led by Ilya Tolstoy, grandson of the Russian author Leo Tolstoy, and several conservationists from The Bahamas and the United States. They concluded that the fragile ecologies of the islands were under threat unless something was done to conserve them.

The BNT is responsible for the upkeep of national parks, including the Exuma Cays Land and Sea Park. Its Exuma headquarters are on the island of Warderick

Figure 58. The Exuma Expedition. (Courtesy, Bahamas National Trust)

Wells Cay in the park, where cruise-goers pay the fees that are required to enter the park. Although the BNT receives an annual subvention from the government of The Bahamas, most of its income is derived from donations, which often, we were told, come from "wealthy white people." These wealthy individuals were often expatriates attracted by The Bahamas' liberal tax regimes who might live in New Providence, Exuma, or the other islands, and not necessarily the super-wealthy private island owners. Other forms of fundraising come from the annual weekend-long Jollification, a party that marks the start of the holiday season and takes place the weekend before the American Thanksgiving holiday in November. Jollification takes place at the BNT's headquarters on New Providence, called The Retreat. Surrounded by trees, the Retreat feels removed from the encroaching city. Our engagement with the NGOs and government came largely through executive education courses headed by Paul Nakazawa, associate professor of the practice of architecture. Also joining the courses was Padraic Kelly of Buro Happold in London. These annual courses were designed to test the work of the Sustainable Exuma project and to generate discussions on the initial findings with members of the government and NGOs such as the BNT.

Churches are another important informal form of local governance in Exuma. The Exuma Cays are firmly Baptist, although there is at least one proud Muslim on Little Farmer's Cay. George Town, on Great Exuma, has the greatest number and range of churches, including St. Teresa's Catholic Church, St. Andrew's Anglican Church, a Seventh-Day Adventist Church, and several smaller Baptist and Evangelical churches. Nearby Stocking Island even has a "Church on the Beach" for boaters. We tried to have researchers attending as many churches as possible, as we found it was a good way to meet a broad section of society. We noticed a colonial holdover among the churches with the Anglican, established church, at the top of the hierarchy, Baptists in the middle, and Catholics at the bottom. Many of the immigrants we encountered were Catholic.

Fourth, we engaged with tourists, on whom the economy of Exuma is dependent. A sizeable percentage of tourists return every year and have become part of the local community. At the Peace and Plenty Hotel in George Town, on Great Exuma, for example, one room is named after Peter and Ingrid, a German couple who returned every year. Most tourists, however, don't stay very long or engage deeply with residents; they tend to either spend their days isolated on private boats or in larger resorts. During the high season of January to March, it's not uncommon to see up to five hundred private yachts and sailboats docked across from George Town, forming an aquatic city of their own that at nighttime appears like a constellation of flickering lights. Many of the boaters come from the southern United States. A standard air route brings people into Exuma airport outside George Town, which has direct flights from Miami, Atlanta, Charlotte, Fort Lauderdale, and Toronto. Alternatively, visitors fly into Nassau and catch a smaller plane to Exuma. Once in Exuma, visitors often sail northward along the coast to New Providence, a journey that can comfortably be spread over a week. Along the way, boaters will stop off at institutions such as the Staniel Cay Yacht Club.

In Great Exuma, visitors are served primarily by the local establishment Chat 'n' Chill Beach Boutique, a short boat ride away on Stocking Island, which was founded to alleviate boaters' need to go into town to get supplies. Chat 'n' Chill, whose mantra is "No shoes. Kick back. Just relax. We're on island time," includes the popular Exuma Beach Bar and hosts that aforementioned Church on the Beach on Sunday mornings. Grocery stores, such as the Exuma Market in George Town, often have one set of shelves and prices for locals and another for tourists—a testament not only to their distinct food preferences but also their different economies. Even seemingly basic foodstuffs can be in short supply on the islands. At the hotel in which we regularly stayed, there was a regular shortage of fresh fruit and ice cream. A server remarked how they had fruit and ice cream in the refrigerator, but the owners liked to keep it for themselves. Coffee was always hard to find for similar reasons, so we brought our own.

The fifth constituency we identified is made up of immigrant workers. They come in two main groups, legal and illegal, ostensibly very different but not that different on closer inspection. Immigrants are the hardest of the groups to talk with, since many are understandably reluctant to draw attention to themselves. Often of Haitian descent, they live lives hidden from the majority but provide essential services on the islands, especially in tourism and agriculture. We met Haitians on Sundays at St. Theresa's Catholic Church or in one of the Evangelical churches in Great Exuma. Some people, we found, attend both churches at weekends.[15] In order to work, immigrants depend on an immigration permit. These permits require the sponsorship of a Bahamian citizen and can range from $500 for farm laborers to $15,000 for a CEO or general manager.[16] As a result, many come illegally, often risking their lives on the sea crossing. We regularly heard and read about tragic boat disasters involving farm workers being brought illegally to Exuma.

Lastly, and importantly, we considered the environment itself as a constituency. We listened to the environment, learned its languages and codes, and treated it as an equal partner for engagement.[17] A planner, someone I admire very much, surprised me with some provocative questions: "How can you study the environment? Do you speak to it? Does it answer you back?" The planner suggested we talk with businesses instead, since they have a direct agency over the land and a direct voice in how it is shaped. When I repeated this interaction to a wealthy billionaire, not from The Bahamas, the billionaire retorted, "But the environment has the largest voice of all!" As mentioned earlier, much of the land on these islands is expected to be submerged in the next thirty years. We took this rise of the sea levels, as well as the frequent hurricanes that batter the islands, as one way in which the environment was speaking to us. Yet, at the time we were doing our research, climate change was not on the minds of many of the locals. And if it was on people's minds, we did not see evidence of what anyone was doing about it. We did, however, bring up the topic repeatedly with the government and the BNT, who were sympathetic but also somewhat powerless in the face of such rapidly rising sea levels and the strength and frequency of tropical storms.

We reached out to each of the constituencies for engagement in different ways. We spent most time with the communities but did our best to engage with the others too. As part of our preparations for fieldwork, we gathered maps, plans, images, and documents that help make the islands distinctive. These helped us formulate the questions around which we based the project.

Guiding Questions for Collective Fieldwork

In the project, we asked four related questions, three of which emerged from the discussions at the project's outset as well as the discussions with government representatives and the Bahamas National Trust. The last question was added toward the end when it became clear we needed to find better ways to communicate. They are

1. How can we ground the project by incorporating fieldwork into the design and planning process?
2. How can we rethink a plan as a process, rather than as a singular document or report?
3. How can we build more sustainably on the islands?
4. How can we communicate better with our audiences?

While these questions have different epistemological natures, they are all related. The groundedness provided by the fieldwork helped us devise a method for rethinking the plan as a process, which led directly to the question of how to build more sustainably. The communication issue was key—not just in terms of learning from the communities, but in terms of how to give something back.

In the following section, I review the process of collective fieldwork; the benefits of a process versus plan, explaining why we went with an iterative method instead of a single plan at the end; some of the highs and lows of fieldwork and design; as well as describing some of our interventions, prescribing a different future.

Breaking Down Collective Fieldwork

After my frustrations with long-term ethnographic fieldwork, described in chapter 3, I knew that I wanted to develop quicker and more experimental ways of doing fieldwork. As I learned in Bahrain, the ethnographic process is not only very lonely, but it takes so long! And the world moves faster than ethnography does. Influenced by my background in landscape architecture and my innate desire to build projects and effect change, I had come to question the assumption that research had to be done in this way. After all, what was the point if the knowledge we had revealed could not be activated? Decisions get made in the real world, and while anthropology is, and has been, closely linked with various political projects, anthropologists often do not capitalize on their agency to effect change precisely because of this disjunction with time.[18] The project in The Bahamas was one such attempt to address the temporal and spatial limitations of ethnographic fieldwork by sharing research responsibility among a collective. It seemed to me that the only way of doing ethnographic fieldwork more quickly was to remove it from the individual and share it among a group. The project allowed me to pose a simple but radical question: Rather than one landscape fieldworker spending a year in the field, what if fifty-two fieldworkers spent a week each to field research? Would the collective fieldwork add up to something equivalent, or more, than a single person spending a year in the field?

There are some precedents for this kind of collective fieldwork in recent years and since the 1960s in particular. One historic example is the Harvard Irish Mission from the 1930s, when a team of anthropologists from across the subdisciplines studied the Irish countryside from multiple perspectives, resulting in publications such as Conrad M. Arensberg's *The Irish Countryman.*[19] Arensberg's book quickly became an anthropological classic, and Arensberg went on to become president of the society of applied anthropology. Of most relevance to the Sustainable Exuma project, and required reading for each of our researchers, was Joel Savishinsky's *Strangers No More: Anthropological Studies of Cat Island, The Bahamas.*[20] Savishinsky, who had participated in the Harvard-Cornell Archaeological Expedition to Sardis, Turkey, as a physical anthropologist, helped run an ethnographic field school on Cat Island in the 1970s under the auspices of Ithaca College. With his students, Savishinsky studied various topics such as slash-and-burn farming practices, bush medicine, boatbuilding, gender roles, and spiritual life on Cat Island. More contemporary examples of collaborative fieldwork can be found in Dominic Boyer and George Marcus's coedited book *Collaborative Anthropology Today: A Collection of Exceptions.* Boyer and Marcus, who published a trilogy of books expanding ethnographic norms, juxtapose several examples of collaborative anthropology to explore what they term as a recent surge of interest in new kinds of ethnographic and theoretical partnerships.[21] It's worth noting that architecture and design underwent a similar process of diffusion in the past hundred years or so. Architecture was usually the work of one grand maestro and their minions, and today most architectural practice, even that of *starchitects,* is a much more collective effort.[22]

The Sustainable Exuma fieldwork differed from most forms of collaborative fieldwork in two significant ways. For one, while typically individual authors still write up their own research even within a collective project, in this case each of our authors became part of a single collective voice. Second, the Exuma project was future-oriented. While every student was required to observe and participate, they also had to propose a design or policy intervention for the archipelago. The design or policy intervention was not just to be informed by a student's own fieldwork but by that of the collective. Could we, through a collective endeavor, simulate some of the advantages of ethnographic fieldwork while limiting the disadvantages? Was it possible, I wondered, for collective fieldwork to have a similar depth and thickness and rigor? Undoubtedly, the results of the fieldwork would be different, but I hoped they might also be stronger because a collective could include many more diverse perspectives and a more balanced view of the site than a single fieldworker could capture.[23] At the same time, it is common to have collaborative projects in a design school, but our emphasis on the ethnographic training provided a common methodological basis on which to work. The Sustainable Exuma fieldwork sits alongside forms of community engaged design research as a complement, and the immersive ethnographic experience differentiates it.

After much internal debate, we purposefully did not call the process "ethnography." I came to realize that a distinct form of immersive fieldwork was emerging that I termed "landscape fieldwork." The Exuma fieldwork processes adapted ethnographic methods to be better integrated within design and decision-making in the built environment. By delegating fieldwork among many researchers, we found the project could be more realistically integrated within the timeframe of a design and policy-planning process, given the four- to five-year cycle of government turnover. While the act of collaboration was relatively straightforward, the interpretation of the fieldwork, and the curation of the results, was more complicated when that responsibility was shared among a group.[24] This curation of the fieldwork was to be one of the main challenges.

Our long-term engagement with the communities was complemented by regular meetings with the government and institutional partners, both in The Bahamas and in Cambridge, Massachusetts. The fieldwork was done under the auspices of the "Design Anthropology" course I had developed along with Steven Caton, professor of anthropology. The class was split evenly between students at the Graduate School of Design (consisting of architects, landscape architects, urban designers, and planners as well as master's in design studies students) and students of anthropology, at both the undergraduate and doctoral level. For the first year of fieldwork, our students worked in groups of two: one designer and one anthropologist. They lived and worked together for a week at a time doing fieldwork. Steven Caton and I visited each of the pairs in situ. Then, in the second year, we formed a collaboration with Professor Nicolette Bethel, of the College of The Bahamas, now the University of The Bahamas. Professor Bethel offered a parallel course in Nassau. In the second and third years of the project, when Professor Caton stepped back and I offered the course alone, our students and researchers went to the field in groups of three, typically one from the Graduate School of Design, one from the anthropology department, and then one Bahamian student, usually from the fields of sociology or journalism. We assigned students to islands based on their connections. Many Bahamians, we found, had cousins or friends on the islands who would act as hosts, introducers, and interlocutors. Our students were usually assigned at random, or else we decided where each person would go according to some quirk or trait. For instance, Alex from Long Island, New York, went to Long Island, The Bahamas. We felt that being from the New York equivalent might be a good conversation opener, and it was. Shahab, from Kuwait, went to Little Farmer's Cay because we thought he might get along with Mr. Bevans, a resident from the island who had converted to Islam in his twenties and remained a fervent Muslim. As we hoped, they got along together splendidly. Sunny went to Staniel Cay because of her interest in water, waste, and recycling and the controversial desalination plants and incinerator there. Sadie went to Cat Island because of her interest in death rituals; a senior figure in the government had told us that no one dies a natural death on Cat Island. (We presumed he was referring to Obeah, a spiritual practice derived from enslaved peoples similar to Voodoo and Candomblé, but different in the absence of a pantheon of gods and collective worship.[25] To this day, Obeah is illegal; associated with magic and sorcery, it's assumed to be practiced on remoter islands such as Cat Island.)

So, while the fieldworkers shared an overarching question—to find out about how residents envision the islands developing more sustainably in the future—the students brought their own interests and passions to the fieldwork too. This was an important aspect of the method. In general, they didn't find it hard to find people to talk to. When students were in more remote areas, in general we preferred them to rely on rides from neighbors than to provide them with cars. We encouraged students to find ways to generate chance encounters.

The first six weeks of the course prepared students for fieldwork by introducing them to ethnographic methods by reading what others have done, in The Bahamas and elsewhere. We also had some fieldwork assignments, since by far the best way to learn to do fieldwork is to do it. One of these assignments was to spend one hour in a public space and then four hours interpreting what they would see. This reinforced the ratio between observation and interpretation which is so central to our method.

Immersing themselves in the field, making decisions instantaneously, and collaborating with other fieldworkers were all part of the learning process. The core aim was for fieldworkers to engage various constituencies through the power of conversation, usually in the form of unstructured interviews. A secondary form of engagement, which was nonetheless important, was the conversations our interventions sparked among the interlocutors themselves when they were at home, in the store, or walking home from church. We had no control over these secondary conversations and never really knew their content, but we knew they took place.

Landscape fieldworkers lived in the small communities for about ten days each, usually in mid-March during the college spring break. During the ten days, we expected it would be extremely difficult to get the depth and thickness we needed from fieldwork. Surprisingly, it was almost as if the compression of time forced

Figure 59. Landscape fieldworker in the field with tablet talking and walking with a resident. (Sustainable Exuma)

islanders to not hold anything back. We found they were often incredibly open with our students—admittedly, with some of our students more than others. Of course, this might not be about the students but the residents themselves, and highlights the importance of personal dynamics to fieldwork as much as issues over gender, class, and religion. Residents told us about what they needed on the islands: jobs, a full-time doctor, more tourism, flights to Fort Lauderdale, cheaper energy, more cash on the islands. People told us that they craved fast food: they wanted a Dairy Queen or another fast-food restaurant to come in. To this end, barter systems exist whereby Exumians send fresh fish to family members and friends in Nassau in coolers via the daily flights on Flamingo Air, Air Bahamas, or Pineapple Air. In return, islanders receive coolers filled with KFC (Kentucky Fried Chicken), Big Macs, or ice cream from Dairy Queen. This phenomenon was clearly visible every time we took a flight.

Many Bahamians told us that they felt "overstudied." They told us that researchers come into The Bahamas, but the research leaves the country and doesn't return, and they resent this. We heard this on the islands, we heard it from government officials, and we heard it from the prime minister. Because this project lasted three years, and we revisited the same communities, in the eyes of many community members, the students returned. They treated the fieldworkers as the same individuals even though they were different people. Some residents didn't realize the persons had changed; they saw them as the "Harvard and COB students." They saw the new faces as old friends. This aspect of return, it transpired, was important for building trust between fieldworkers and interlocutors. A resident of Little Farmer's Cay told me that this was the first time they felt anyone "cared enough" to return. They remarked on how unusual it was for planners and designers not only to sleep on the island but to come back. For all sorts of reasons, practical and economic, few designers can live among the communities they are designing for. Z+T Studio, in Shanghai, is an exception; members of the firm have been known to live among villagers to better understand both the villagers' needs but also their skills and construction abilities.[26] One of the principals, Ziying Tang, trained in human geography before landscape architecture.

In many ways, ethnography is the painting of a portrait. Like David Hockney's portraits of his mother, made up of multiple Polaroids, you can sometimes get a better picture with many images rather than just one. Ethnographic fieldwork has a similar agency in terms of thick description, looking at an issue from many perspectives. However, the multiple Polaroids that comprise Hockney's portrait are taken by the same person. If we were, however, to imagine this collection of Polaroids with each one taken by a different artist, that would be the visualization of collaborative fieldwork: many viewpoints from different perspectives of the same, or similar, subject.

We required fieldworkers to write very detailed notes based on the guidance I follow for my own research: for every hour in the field, we need four hours to interpret it through writing, diagraming, drawing, and sketching. In addition to noting what people said, we asked students to describe the persons they interacted with, the color of their eyes, their gestures, and so on. All of this is guided by Clifford Geertz's famous distinction between a twitch and a wink. Geertz looks for cues to interpret whether the blink of an eye is intentional or not, and thus whether an involuntary twitch or a deliberate wink.[27] And we asked students to look for such details and for patterns as they engaged in everyday life by reading and rereading their notes every day. One of the main logistical challenges we faced—apart from the logistics of sending students and researchers to various islands—was that in ten days, we would gather about a quarter of a million words in fieldnotes.

Creating "Exuma Topics"

As the project went on, we had to figure out how to manage and curate the copious data we were collecting. In a week, eighteen researchers might hold anywhere between two hundred to three hundred conversations with inhabitants. And in a population of seven thousand, that's touching on quite a large percentage of lives. In the end, we had about a million words to analyze from all the fieldnotes collected. We first developed a system of index cards, where two to three specially trained content curators took highlights from the fieldnotes. We then paired these quotes with an image. Images might have been taken from the collective photo archive, or from one of the watercolors students painted, or sketches they made *in situ*. The images may have been a diagram that was constructed specially to illustrate some point or data in the quote. The image acted as a graphic headline. The juxtaposition of image and text was important, sometimes throwing up new associations. This was inspired

NASSAU
25° 3' 36.0" N / 77° 20' 42.0" W
3.2013

-no title-

Architecture in Exuma represents the diverse and dynamic cultures that intersect in the archipeligo

Also in Williams Town, but almost at the back or the settlement, there was an old building that, according to the indications, was part of the old plantation. Was struck me was that, if it was so, the building was completely abandoned. It was supported by stonewalls covered by plaster, supporting a beautiful wood structure on top. The walls were certainly not built on brick or blocks, but from milestone. It seemed to me that, even clearly being a colonial building from an English tradition, it was the closer typology to traditional Bahamian house, than wood houses. One might argue that it would be necessary to look at Lucayan architecture, but regarding that Bahamians are mostly descent of slaves and this kind of building would the kind of building were slaves would live, then this would be the building accompanying the memories

RDU002 / JCO022

Colonialism,Architecture,Slavery,Heritage Sites

Figure 60. Nassau index card. (Sustainable Exuma)

by my prior solitary fieldwork in Bahrain (chapter 3), where I was using index cards to trace key themes running through my notes. From this process we identified key topics on the minds of Exumians. The identity of the author of the text and image were coded on the card, and keywords provided a summary of key topics of conversation.

In Bahrain, I sorted out the index cards on the floor of my apartment. With the Exuma project, the threads were traced digitally through a specially developed platform that we called Exuma Topics. Working with the content curators, we developed an online system for indexing this material that allowed us to tie together key insights and topics that were coming up in our conversations. This online system was very similar to the Instagram online platform. It contained an image, a short text (a "tweetable quote," i.e., with a word limit), keywords, and the reference to the author of the text and image—who were usually not the same person. The online system kept track of keywords like agriculture, waste, fishing, education, mobility, and so on. We found, for example, that climate change and sea-level rise was not a topic that ever came up explicitly, despite its urgency and visibility. But climate change did come up indirectly, such as through concern over the increased number and severity of hurricanes that were taking place. People were noticing that bird patterns were changing and that there was increased flooding in Exuma. They knew that there was not much agriculture in Exuma anymore because the soil was blowing away with hurricanes. This was an especially important topic, since we knew that landscape architecture was fundamental in the response for dealing with hurricanes, rebuilding soils, and dealing with extreme climate.

LFV001 / LFV002

LITTLE FARMERS CAY
23° 57' 27.96" N / 76° 19' 20.25" W
3.2013

Airport Multi-Use

Like docks, the runways are an interface for the flow of people and material from off island, creating a space for a peculiar sociality.

The airport receives a private plane several times per week. However, as urban space it is not what we would expect from a traditional airport. "Best lunch is served at the airport restaurant on Sundays" a fisherman told me. A long line of concreted pavement in one border of the island, it has two buildings by its side; one restaurant and one waiting room. Parallel to the concrete line it is what the pastor advised to be "The best beach of Little Farmer's Cay" "I don't go to other beaches because I don't like the ocean." The airport not only brings people, but also it brings packages that relatives that live in Nassau send to the people from the Cay.

Runways,Infrastructures,Mobility,Diaspora & Emigration

Figure 61. Little Farmer's Cay index card. (Sustainable Exuma)

Figure 62. Exuma topics in exhibition at the National Art Gallery of the Bahamas, 2016. (Marisa Wells)

The Limits of Public Meetings

In addition to the fieldwork among the communities, we held a series of public meetings across the islands. However, it turned out these formal meetings were not as useful as participant observation for helping us learn about the communities we were designing for. The purpose of these meetings was to introduce ourselves and the aims of the project and, most importantly, to listen. We wanted to know what the residents' hopes and ambitions for the future were. At one public meeting in Exuma, someone put up their hand at the beginning and complained, "The light is not working on the pier." We replied with something like, "Thank you very much for telling us this, but we can't fix the light. We're here to hear about your hopes and ambitions for the future of the island." But the discussion kept coming back to the light on the pier not working. This discussion had nothing to do with the light on the pier; it was a veiled way to address a longtime family disagreement.

This often happens in public meetings (as the other example in my hometown showed). The political scientist Morris Fiorina points out that public consultation often gets hijacked by extreme views, and is not necessarily representative of the public, or of good governance. Fiorina writes: "Much of the debate on civic engagement implicitly presumes that it is a good: the more of it there is, the better off we are. I have argued that such an assumption is invalid, at least in the political realm."[28] It was only after the meeting that a politician who had sat silently through the meeting, took us aside and said, "That's why I *never* have public meetings!" He told us that the person responsible for fixing the light was in the room, and that the many people complaining about the light were relatives, and they hadn't spoken for thirty years. I understood at that moment why the politician had sat beside the door. In the end, we found that a process of embodied engagement in daily life arguably provided a more accurate view of people's thoughts and ambitions. We saw landscape fieldwork as a useful complement to public meetings.

As the project moved out of its initial research phase toward design proposals, the process really caused a deep ontological confusion, or disorientation, since we were at once observers and designers. Keeping notes on the design decisions being made challenges designers to really question themselves and their motives. It was an exhausting process of self-reflection. But when we do these projects, we keep notes too. This is where it gets complicated. Not just timewise but because you are struggling to get a job done in a limited time.

Benefits of Process versus Plan

The single issue that the various institutional partners agreed on when we started this project was that no one wanted a plan that would "sit on a shelf." The government and the Bahamas National Trust described a pattern in The Bahamas—which is common all over the world—of a land-use plan being prepared every four or five years. The plan would be commissioned by a newly elected government. A plan would be submitted to the government after a couple of years. It would sit on a shelf. The government representatives would change after a four- or five-year cycle. The new government would not want to implement the previous government's recommendations. It would sit on a shelf. And then the cycle would begin again. At the start of this project, the government did change, which held up the project at first.

However, we managed to break the cycle; the new government supported the project, telling us, "There's just one government in The Bahamas and we support the project."[29]

The idea was that we would initiate a series of smaller initiatives during the larger project, so the chances of the plan getting caught up in bureaucratic processes were minimized. Rather than having one, fixed document at the end, the plan that we completed was already in-process. At the conference that launched the multiyear study, Mohsen Mostafavi, then dean of the Harvard Graduate School of Design and my co–principal investigator, described our intentions: "This is a project that we see as having multiple moments of impact, multiple moments of realization as opposed to a multi-year project that then has a conclusion at the end."[30] If the overall project consisted of an accumulation of several smaller projects, then the eventual plan we submitted at the end would already be under way and, therefore, could not sit on a shelf. It became clear early on that to enable such an active form of engagement, we needed to be in the field. We needed to have an active form of participation with the users.

Fieldwork was central to this form of practice because it provided the insider knowledge we needed to gain the necessary insights for the smaller projects and to initiate them. We found we had to deal with chance and serendipity, something that I had learned to cultivate in Bahrain. In The Bahamas it seemed to just happen. Celebrating chance is common in landscape fieldwork, but the Exuma situation was exaggerated on account of the islands. It was hard to make plans, and when we did, they rarely worked out as planned. As a result, when opportunities came along, we had to be able to identify the opportunity and learn to move quickly. The chicken coops we later designed and built are an example of this active process of engagement.

The chicken coops are a good example of how our active process of engagement met with chance and serendipity. They came about because Rob, one of our team members based in Cambridge, happened to be texting one day with his friend Kirby, who was stationed in The Bahamas for training with the U.S. Navy. Kirby explained that as part of the training, they had to do community engagement. We immediately came to an arrangement with the U.S. Navy's Seabees: we would design something, and they would build it. But we needed to act quickly. It was then that we combined the observations from the various islands into the manifestation of a chicken coop. We designed two types of chicken coops, one large and one small. We selected three different islands where we built the coops. (I got into trouble with Harvard for buying thirty chickens, as, understandably, faculty are not supposed to buy live chickens on a corporate card. I had to sign a declaration that they were going to be treated well, and we did our best to ensure that.) This very small intervention had much larger spatial reactions, as other nearby communities started contacting us via Facebook and email to ask for chicken coops as well. We ended up developing and offering manuals on how to build a chicken coop and how to take care of and raise chickens. The local media thought it was an "eggcellent idea."[31]

The chicken coops were as an example of the cross-scalar project: a small project that embodies a range of regional, national, and global processes.[32] Later, one senior government official told me, "Those chicken coops were revolutionary." I smiled saying, "It's a chicken coop. It's a small thing." He replied, "No, no, I'm very serious." The official explained it was the first time that someone could go into The Bahamas from the outside to build something—with the Nassau government's support—but not going through the deep bureaucracy. And he felt that that was a very significant moment because governance is centralized in Nassau, often to the detriment of the outer islands, and the bureaucracy can take a very long time.

Another example that illustrates the processual nature of our project came during the third year of fieldwork. I was just about to catch a boat from Black Point to Staniel Cay when Caroline, one of the residents we had been working with, came running to the pier. "Come and see my garden," she pleaded. I explained that we needed to catch the boat. "*Da boat'll wait,*" she replied. When we arrived at the holiday cottage, I saw a building surrounded by neatly tended gravel. The gravel was interspersed with rocks and small beds of geraniums. It was not the water-guzzling luscious green landscape that I knew Caroline had yearned for previously. A year earlier, she and one of the students had spoken at length over the aims and ambitions for the landscape around her house. The student had proposed a low-maintenance garden. But Caroline had wanted a green garden with grass and trees. She showed images of villas in Miami as examples of the garden she was imagining. The student explained to Caroline and her husband that it would cost a lot not just to plant the greenery but to maintain it. The couple replied that the cost did not matter. The student explained that the bricks they wanted for the driveway would have to be imported to the island, at great cost. Again, the price was not the issue. The student gave up, realizing Caroline and her husband's values were rooted in a system that appeared to deeply diverge from her own values and interests in conserving scarce resources. A year later, though, it was clear this exchange had had an impact. The couple eventually saw that it was possible to have a garden that they loved that was low-maintenance and cost-effective with a low-carbon footprint. I was happy that day to see the spatial impact of one-to-one conversation.

Some of our field encounters were more fraught—though we tried to see these, too, as learning opportunities that we could use to revise our approaches as we went along. During our fieldwork, everyone concerned with the project went to church every Sunday when they were in Exuma, to meet and connect with the local population. Although we were not necessarily concerned with the theological aspects, the sometimes fiery preaching provided valuable insights into peoples' values and ways of life, and it was an excellent way to strike up conversations. So, when our project collaborated with Belinda Tato and Jose Luis Vallejo from Ecosistema Urbano, a Spanish firm known for their digital forms of community engagement, we decided to plan our intervention after church one Sunday. We greeted churchgoers departing from St. Andrew's Anglican Church, which had a mostly local congregation, and from St. Theresa's Catholic Church, which had a congregation made up mostly of immigrants and tourists.

We asked people to write their dreams and hopes for the future on pieces of red and blue paper.[33] We were interested in the projections of residents, before we would influence them. The congregants wrote what they would like to change on one sheet of paper and one thing Exuma needs on another. All the sheets were then folded into red and blue origami-style flowers and "planted"—just for one day and night—in the main public space in George Town, a space in desperate need of a redesign. After the intervention, some individuals formed a group to consider the more long-term design of this space.

While we considered this temporary garden successful, there was one incident that gave me pause. On a Sunday evening, we had set off in a car with the hopes of meeting people at random with the survey. As we drove through the settlement of Williamstown on Little Exuma near the Tropic of Cancer, we saw one individual

Figure 63. After church on Sunday, St. Andrew's, George Town, Exuma. (Photograph by Emilio P. Doiztua)

Figure 64. Garden of Dreams, George Town, Exuma, a collaboration with ecosistema urbano. (Photograph by Emilio P. Doiztua)

and decided to approach him. At first, I did not realize he was intoxicated, but as we talked that soon became evident. What was more troubling was that he was also in deep emotional distress. It was not the right moment to ask him about his hopes and dreams for the future, but it was too late by the time I realized this. I recognized his hopes and dreams were in emergency mode, probably not extending much beyond that moment. I came face to face with my inability to alleviate his suffering. I found myself asking, "Who gives me, a designer and researcher, the right to ask questions like that? By asking about his hopes and dreams, am I implicitly giving the impression that I can help?" Such encounters raise ethical issues over the role of the landscape fieldworker, and in fact, is why many anthropological fieldworkers are hesitant to do any form of applied fieldwork. We tried to learn from our mistakes by being more careful in our interactions and the contexts in which we approached residents.

While not perfect, these are the kinds of processes that we initiated on the ground. So, rather than submitting a final document that would go to one ministry or government department, we developed a toolbox of ideas and projects aimed at multiple agencies for consideration and implementation.

Fieldwork and Design: Highs and Lows

At the beginning of the three-year project, we made an annual report for our collaborators. The report contained summaries on the various strands of the work, including a symposium we had held at the beginning to discover the issues and set the terms of the study. The first report was a beautifully designed object. While it was an excellent way to communicate with one another and with our colleagues at the university, we soon realized it might not be the most effective way to communicate with our interlocutors. It was sought after by a wide section of people in Exuma and beyond, but not for its content. Its thickness was very good for raising the level of a computer screen or being used as a doorstop.

Later, when the project was over, we created a final draft report of one thousand pages that we submitted to the government. One of our Exumian friends, T.J., took a step back and said: "Don't you realize we have an oral tradition here!" T.J., who has an MBA from the Wharton School at the University of Pennsylvania and runs a boutique resort on one of the islands, insisted that Exumians communicated through conversation and storytelling rather than books. Indeed, the standard reaction among the Exuma Alliance, a group formed to implement our recommendations, was hesitant: one thousand pages was too many. And within these small island communities on Exuma, people's reactions were, unsurprisingly, even more detached. Meanwhile, in Nassau, "We don't have the time to read it!" quipped one of the senior figures in the Ministry of Works. So, we started thinking about how to communicate intricate design and planning issues in ways that are more accessible to the public they are aimed toward.

Considering that The Bahamas has an oral tradition, we started rethinking the methods of communication with the people we were designing for. As a result, we developed an exhibition as a way of handing the project over to the Bahamian nation. We chose the National Art Gallery of The Bahamas, the NAGB, for the exhibition. It was far from ideal, because we feared it would reinforce Nassau-centric hegemonies, but we chose it for its national profile, and proximity to government ministries and the College of The Bahamas (COB). It was close enough to the center of government for us to expect ministers and officials to visit, and the interactive format of the exhibition we designed allowed the COB to hold classes in the exhibition. We intended to travel the exhibition to the individual islands. But the traveling was logistically and politically challenging. Not only did we have the difficulty of moving the exhibition around, but there was the question of where to move it to; we risked alienating some communities by having the exhibition in some places and not others. So, to move it to the islands, it needed to be portable.

Our solution was what we called a "toolbox"—a hybrid booklet/traveling exhibition, allowing the publisher to be the distributor via online networks. The toolbox included nine different tools, formatted as a poster that could be folded out and displayed at various schools and church halls, or even mounted in a public space. These nine tools were also shared with the government. An advantage of having several instruments, rather than one, is that they can be activated by various constituencies at different moments in time. So rather than the entire plan going to, say, the Physical Planning Directorate and sitting on a shelf, the planning component with its proposed codes and regulations is just one of the different tools that we submitted. Projects can be activated by government, NGOs, community groups, private developers, and/or individuals.

Figure 65. College of The Bahamas Exhibition, used as a teaching space. (Photograph by Marisa Wells)

The exhibition forced us to boil down the complex project that we had summarized in one thousand pages into a few memorable propositions. We identified four words that underlay the dynamics of the process: engage, work, share, and imagine. We explained them as:

1. Engage widely: participate in everyday life and look for patterns.
2. Work across multiple scales: consider the very small and the very large, from each person to the nation—this is where landscape architecture can be especially helpful.
3. Share knowledge: disseminate what you've learned through briefings, community meetings, videos, one-on-one conversations, social media. To this extent, we attempted to engage with as broad an audience as possible.
4. Imagine the future: provide images of the future—both preferred and unpreferred futures. If we don't imagine the future, we have no way to shape it.

While we ultimately went with a visual and tactile approach, rather than conveying our results through oral means as T.J. had suggested, we still found the approach a more successful way to communicate and engage audiences than the written word alone. We worked closely with Hoon Kim and his creative consultancy Why Not Smile LLC (WNS), based in New York City, to develop the "exhibition in a box," later termed a "booxhibition."[34] The exhibition-in-a-box was inspired by the practice of *literatura de cordel. Literatura de cordel* (literally, string literature) is a popular practice in northeast Brazil; it consists of displaying unbound booklets in public spaces on a string like a clothesline. In this case, we imagined the unbound booklet and posters being distributed throughout Exuma and mounted in public spaces, church halls, schools, and private homes.

Why Not Smile developed the project's logo, composed of individual dots or islands, adding up to a single whole. One memorable moment came with WNS's design for T-shirts. The shirts, designed with a single dot, were deployed across the islands. They were at once an individual piece of clothing and part of a collective. They were easily recognizable from boats as moving bright blue dots.

One of the projects in the box was called "Grow Fruit," and we developed it in response to the shortage of fresh fruit in Exuma that we encountered. We proposed propagating fruit trees to provide fresh fruit, improve public health, and have a positive visual impact on public space: we had noticed an absence both of fresh fruit on the islands, and of low-growing trees. A year after this idea was first generated in a workshop, someone told us that inspired by the project he had already propagated two thousand fruit trees for planting in Exuma. We attempted to have multiple ways of entry into the project, so that it was not dependent on any one agent to implement it. Rather than have one agency responsible for a mass fruit planting, could many people buy into whichever parts of a project that interested them most? We designed calendars relating to the Grow Fruit and, later, Cook Fruit initiative, inspired by Guattari's *The Three Ecologies.* Designed to inform through pleasure and taste rather than didactic instruction, the calendars were structured around the weeks of the year with a different fruit or recipe per week. We used the calendars as an inspirational and pedagogical device. These calendars were disseminated quickly among the Exuma population and schools, and we had to print more.

The Grow Fruit and Cook Fruit projects were examples that could be extended to other areas, construction, for example. As we learned from the chicken coop process, almost all construction materials on Exuma are imported and expensive. They usually come from another island, and often before that are imported from the United States. How could we activate projects with small island communities with limited resources, even resources that are already on the islands? Inspired by Felix Guattari's *Three Ecologies,* we looked for materials and methods that were environmentally, socially, and aesthetically sustainable.

Anticipating Alternative Futures

If we don't imagine the future, we have no way to shape it. To help combat this, our team developed a series of scenarios of what the future might look like on the islands. We were guided by the anticipatory anthropologist Robert B. Textor.[35] Textor has seven points and tells us, "Our anticipations of the future do, after all, often turn out to be essentially correct." He encourages us to envision a preferred future, saying: "The best way to realize a preferred future is first to visualize it—in concrete enough terms to provide the motivational basis for appropriate action."[36] He makes the point that if an individual or society does not envision the future, they risk being terribly disappointed with the result whenever, as he puts it, "the erstwhile future becomes the present."[37]

Sea levels are rising, and the Bahamian populations are increasing. What will happen in the future? To enter the present and future space of the islands, we imagined scenarios, often taking extreme cases. The logic being, that if we are prepared for an extreme case, then the reality will likely be less extreme. These scenarios were informed by fieldwork as much as by our own skills as landscape architects. At the time of writing, about 70 percent of the population lives in New Providence, which is nearing capacity, and just 2 percent live in Exuma. So, the Bahamas faces a decision: They can plan for increasing density in New Providence, which would likely mean expanding upward with high-rise buildings, though these might incur damage from hurricanes. Or they can decentralize on the islands and develop relatively "unspoiled" areas like Exuma. Given the likelihood of the latter option, we developed a planning framework to help protect the coastline of Exuma from development as well as rising sea levels. We simulated projections for the islands, as many of the areas where people are living today will be under water in the future.

We also considered how we might positively embrace such changes. Based on our findings, it is not so fanciful to imagine a future where people might live and farm in the sea, revive salt collection, and imagine new ways to collect water. Such a future will likely see a more decentralized Bahamas, one that is less Nassau-centric, and one more focused on the Family Islands, including Exuma. We, for instance, proposed a strategy of planting coconut palms along the seafront. They would be important partners in coastal remediation and the straw used for local industries; we call it a "strawtegy."

Landscape architects are very good at working across scales. Our proposals ranged from the small-scale chicken coops to larger-scale strawtegy. This multiscalar ability is of particular importance when dealing with a large and fragmented site composed of small island communities where it is essential to understand not just how to work on one island but how to work across them. How could we begin to grasp some of the complexities of the site, given that we were dealing with 365 islands over 120 miles (190 kilometers), with many different communities and landowners? To do so required a shift of focus, not just from one island to all islands, but to the scale of the individual to the islands to the scale of the globe, and vice versa. This involved an acknowledgment that the role of the landscape architect is to design more than space; we are also responsible for the effects of what we do.

To aid with these scenarios, we developed a series of GIS maps for Exuma. As well as a series of thematic maps on a range of topics, such as island geology and ecologies, water and climate factors, development frameworks, and land-use guidelines. We exhibited the draft maps in among communities in Exuma for their feedback and critique. We also drafted an atlas for the islands.[38] We also developed a set of panoramas, taken from boats, recording the coastline in the face of rising sea levels. This was also the perspective from which many people see the islands: from the sea looking in toward the coast. The various layers differed from the layer-cake method developed by Ian McHarg because we introduced the human dimension, challenging the positivist deconstruction that McHarg proposed. We traced not just the physical characteristics of the islands, but their social ecologies too.

Between the National and the Local

It is also clear that there is a level of governance missing in The Bahamas, in between the nation and the community. We need people to talk to about chicken coops on a regional scale: regions are missing from the level of governance. The landscape ecologist Richard T. T. Forman tells us that regions are "a group of landscapes."[39] These landscapes are made up of local ecosystems.[40] Forman defines a region as "a broad geographical area with a common macroclimate and sphere of human activity and interest."[41] He goes on to stress that concepts of regionalism link the physical landscape with environmental processes and political and social structures. One can begin to imagine groups of islands that have a regional capacity that is tied to territory, rather than population. This sort of subdivision is essential for the proper management of the land.

Our work was partly a reaction to Carl Steinitz's "From Project to Global: On Planning and Scale," which I had first heard as a riveting keynote address at a conference in New Zealand.[42] The content of the lecture was compelling, although I did not agree with everything Steinitz said. In discussing his concept of geodesign, Steinitz quotes Galileo: "Many devices which work on a small scale do not work on a large scale."[43] He makes the point that you cannot do everything: you must let go of the large, or the small scale; "Or you must let go of depth, time or experience."[44] In this project, we included smaller-scale designed projects as prototypes within the larger regional framework.

We were also inspired by a quotation from the anthropologist Margaret Mead, who in an interview with the landscape architect Ian L. McHarg asked: "Do you, when a hurricane comes up, run around with sticks in your hand and beat something hard enough to make more noise than the storm and scare it away? Or do you climb a tree with a knife between your teeth to cut the fingers off the wind?"[45] We did not try to cut the fingers off the wind when a hurricane comes, but we tried to learn about traditional forms of knowledge, as well as skills, hopes, dreams, and ambitions as an essential part of the process that we developed. For example, a senior figure in the ministry told me how she did not trust the weather forecast. Instead, she trusted her father's method of predicting hurricanes handed down on the islands over generations. This involves licking your index finger and stretching out to sense the direction of the wind. She maintained this could predict the direction of a hurricane with more accuracy and with more lead time, so there is time to prepare. We were interested in such forms of so-called indigenous knowledge in parallel with the more conventional and scientific knowledge that we were trained to work with as designers.

Landscape fieldwork helped us to see that many of the skills needed for the future are already in Exuma. They simply are not being shared between generations or between different social groups. We asked ourselves how we could use new media, including social media, to share this knowledge and aid in communication and fact-finding. The visual anthropologists and filmmakers Siri Linn Brandsøy and Suneeta Rani Gill made a series of videos of skills on the islands called "For the Future of Exuma."[46] These remarkable short films, with additional camerawork by Tarjei Langeland, record details of everyday life from beekeeping on Great Exuma to fishing in the Exuma Cays. These everyday interactions with land and seascape are often not shared between generations any more, so the films provided a way to document these forms of knowledge. They did so by capturing the sounds, accents, and images of the islands as much as the skills the interlocutors were discussing.[47]

Future Considerations

Landscape fieldworking collectively presents numerous advantages in terms of efficiencies of time and space. Collective fieldwork is a useful way to gain the pulse of a community. It could be likened to a triage nurse (or a team of medical personnel) taking the vital signs of a patient in an emergency room, as opposed to a singular doctor doing a full medical examination and diagnosis. But is it something less, or more? Perhaps the emergency personnel have a set of skills, and colleagues, at their disposal that adds up to something more than a one-on-one consultation with a doctor. The Colombian landscape architect Diana Weiner sees fieldwork as a basis for a diagnosis of the landscape. She tells us, "If we don't have a good diagnosis of the place with the people, we don't have a base for the design."[48] Indeed, collective fieldwork may be one of the better ways in which fieldwork can be integrated within a design and planning process—especially across a large landscape with multiple constituencies for engagement. The perspectives that come from the collective—

which by definition has more than one—adds diversity by offering multiple perspectives rather than that of a single fieldworker. We proposed that this method of community engagement be enshrined in Bahamian planning practice. In this case, we worked with students within the context of a course. But it doesn't necessarily have to be so; there are other ways to do collective fieldwork. The landscape architect Jala Makhzoumi, for instance, describes how she enlists the help of a community she is working with to gather the necessary data on themselves.[49]

At the same time, a major hindrance, indeed contradiction, is that the collective work still needs a single curator, or curators. And the curators need to know not just the site being studied, but the landscape fieldworker too. With planning and foresight, the curatorial role can be integrated into the design and planning process from the beginning, but it nevertheless requires time. Moreover, the interpretation of the

Figure 66. Farm in the sea. (Sustainable Exuma)

Figure 67. Live in the sea. (Sustainable Exuma)

collective fieldwork is more complicated on account of the multiple voices. To this end, it is essential for the curator to spend time with the individual fieldworkers in the field. This way, they know the context better and can more successfully engage with the process of interpretation. Based on our experience in Exuma, while the gathering of data is quicker, the interpretation of that data takes a lot longer.

One of the biggest surprises we learned from fieldwork was that Exumians, if not Bahamians in general, have a very particular view of landscape and landscape architecture that differs from a more standard understanding of the field. When I would introduce myself as a landscape architect, Bahamians usually assumed that my role was related to agriculture and food production rather than about beautification, for want of a better word, or aesthetics. Occasionally, Bahamians associated landscape architects with ports and government buildings, even gated communities, but more often than not, they thought of us as farmers. I make the distinction of, Bahamian, because non-Bahamians seemed to have a different reaction—perhaps reinforcing the foreignness of the field and their familiarity with the field. Perhaps this association with agriculture could be tied to being on small islands, where concerns over a shortage of fresh food and vegetables far outweigh aesthetics. It could also be linked to the absence of a need for public parks when one is surrounded by commonage and beaches of incredible beauty. It could be linked to the comment I made earlier about the limitations of working in a beautiful landscape. It could also be the limited use of the word "landscape" in the Bahamian dialect. But perhaps the greatest capacity to help the island communities, and to effect change, is from the perspective of

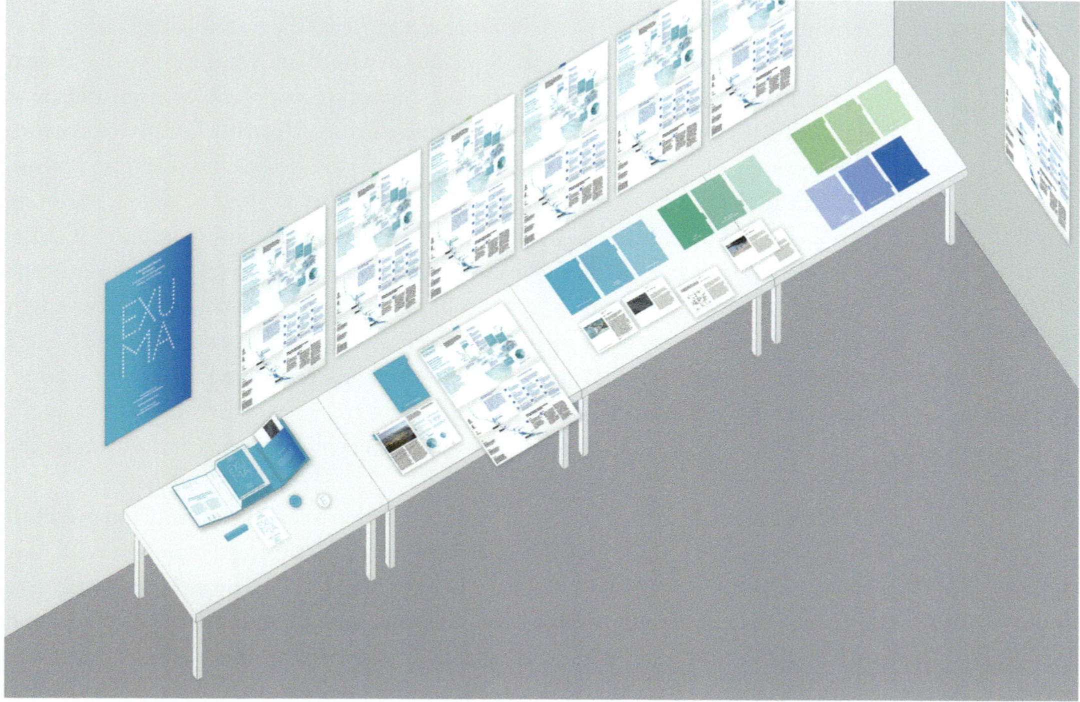

Figure 68. Exuma Toolbox. (Why Not Smile LLC, New York)

horticulture and farming. The Exuma population at large appears to consider landscape architects as farmers. Therefore, many of our projects focused on fresh food production both vegetal and animal. If our Exuma experience can offer general lessons to the field, it is that landscape architects should be as open to growing chickens and fruit as much as planting flowers and trees.

Reflections

In this chapter I have introduced a process of collective fieldwork where the ethnographic responsibility is shared among a collective. The result is not necessarily a collective ethnography; I prefer to see it as one way of doing landscape fieldwork among more than one person. We developed a form of collaborative fieldwork, sharing the responsibilities among fifty-two or more researchers. The Sustainable Exuma project set up a framework for a mode of practice in which landscape fieldwork is linked with both descriptions and prescriptions. In doing so, we debunked any notion that fieldwork is just about observation and description. This collaborative fieldwork allowed multiple perspectives to be integrated into the design and planning process. It also allows the fieldworkers to support one another in their fieldwork, delaying the "fieldwork blues." The fieldwork resulted in an overwhelming amount of information, a snapshot of life in Exuma at moments in time. The core of the landscape fieldwork was the multitude of conversations that were conducted over the length and breadth of the archipelago. These conversations, as in the case of Caroline, were part of the design process, and had multiple moments of impact, mostly undocumented and unmeasurable. The conversations were as real and important as the mini projects we developed and initiated, whether chicken coops or the overarching Exuma toolbox.

Recognizing different epistemological origins between description and prescription is important: landscape fieldwork does not necessarily lead to design in a linear way, but landscape fieldwork can inform design and lead to propositions that have a greater chance of being impactful because they are embedded in the ecologies and needs of communities, and governments. Those design propositions can emerge from the fieldwork or land like a spaceship out of the heavens. By cultivating the rich space between observing and doing, meanings and facts, reflection and action, such forms of landscape fieldwork can uncover varied ways of understanding landscapes, settlements, and islands. Through the four propositions of engaging widely, working across multiple scales, sharing knowledge, and imagining futures, the Sustainable Future for Exuma project established a method for an ongoing creative engagement that is deeply rooted in landscape fieldwork.

And if I don't get it done on Sat'day, it can damn well wait 'til Monday.
So when you're rushin' around, trying to get downtown,
remember, you ain't hurryin' me.

—Eric Minns, "A Bahamian Calypso"

5

GATHERING LEAVES

Urban Ecologies of Afro-Brazilian Sacred Groves

A fruta só dá no seu tempo. (Fruits only come in their own time.)
—Mãe Stella de Oxóssi, *Meu tempo é agora* (My time is now)

It was almost 1 o'clock in the morning, and I was waiting beside the gate for a car to pick me up after a festival at the Casa de Oxóssi, a Candomblé *terreiro* in Salvador da Bahia, Brazil. Before going any further, let me explain some terms. Candomblé is an Afro-Brazilian religion that fuses West African religious practice with elements of Christianity, in particular Catholicism.[1] The *terreiros* are a combination of urban farm and monastery, perhaps even gardens, where elements of nature associated with Afro-Brazilian religions are cultivated through combinations of plants and sacred leaves together with animal sacrifice, religious rituals, dance, and trance. *Terreiros* provide a home to the orishas (from the Yoruba *òrìṣà*), who can be understood as gods, deities, or energies of nature. Indeed, the terminology for orishas is contested. Some suggest that calling them anything less than a god or deity is demeaning and racist. Here, I will refer to orishas, or energies, because those are the terms—*orixá* and *energia* in Portuguese—that I heard my interlocutors using during my fieldwork.[2] Finally, the Casa de Oxóssi is the house, yard, or community, of Oxóssi (Ọ̀ṣọ́ọ̀sì in Brazilian Yoruba), the orisha, or energy, of hunting, the forest, the arts, and beautiful things.

Now, let's get back to the story. A senior member of the community, the Iyakekerê (from the Brazilian Yoruba, *Ìyákékeré*), approached me and asked me to move away from the large *iyami* tree beside the gate, a jackfruit (*Artocarpus heterophyllus*). *Iyami* means "my mother" in Yoruba and refers to the power of older women. The tree was decorated with ribbons, and offerings of food, candles, and sacred objects were sitting around its base in honor of Ossayin, the orisha of leaves. I walked about six meters, or twenty feet, away. The elder said that was fine, I could stop there. But why did she ask me to move? Although I didn't inquire at the time, I assumed then that

Figure 69. Jak or jack tree (*Artocarpus heterophyllus Lam.*), with man collecting the fallen fruit. (Engraving by J. Storer after J. Forbes, 1767, Iconographic Collections)

she was worried that I may be affecting the energy of the tree. Later, it was explained to me by Mateus, with whom I was collaborating in the fieldwork, that the elder was possibly concerned that the tree was affecting me. And why was six meters enough? Maybe the gesture of moving at all was sufficient? When I next returned to the *terreiro* and retraced my movements from that night, I could see that I had moved just outside the tree's canopy. Had the elder assumed when I was no longer under the cover of the tree that I was no longer at risk of being entranced by it?

Entering trance is a central aspect of Afro-Brazilian religions such as Candomblé.[3] Trance takes place during ceremonies when participants who have "the gift" begin to enter an altered state of being at the moment when the orishas "descend." Trance is induced by rituals, music, and material and spatial conditions. Some describe it as spirit possession. Others say it is the embodiment of the self with the orishas. Some say that nothing enters the body, but that trance wells up from the inside when the body resonates with the energies of nature.[4] Whatever the explanation of trance, it is considered a sacred state of being. For some observers, witnessing the onset of trance can be a frightening experience, as any encounter with the unfamiliar. It can come suddenly, or more gently, depending on the orisha. For many tourists, the ceremonies are dazzling and spectacular events that are high on the to-do list when visiting the state of Bahia.

Explaining trance is not the focus of this chapter, though, as readers will see, it is pivotal. That embodied experience, among others, taught me that trance *should* be of interest to landscape architects and other designers. According to some practitioners of Afro-Brazilian religions, trance is only possible due to an extreme potentialization of "energies of nature." I suggest that landscape architects should be open to the idea of a tree having an energy or spirit as much as the tree having a canopy, a root span, a physical form, an intelligence, or an environmental footprint.[5] Landscape architects might also be receptive to the idea that entrancement is cultivated through certain material and spatial configurations. Afro-Brazilian spaces are heavily designed spaces, just not designed in a formal, Cartesian sense. The principles on which they are shaped are handed down over generations through detailed observation and experience. The wider literature on Afro-Brazilian religions tends to focus on the sociological and spiritual aspects of the practice, often fetishizing these trance-like states that some members encounter during the rituals. The aim of this chapter,

Figure 70. Candomblé ceremony, Salvador da Bahia. (Photograph by Pierre Verger, reproduced courtesy of Fundação Pierre Fatumbi Verger)

by contrast, is to explore some of the spatial and material design of Afro-Brazilian spaces and in particular lines of inquiries that can be best experienced and understood *in* the field. There are forms of experiential knowledge that are only achieved through a direct, sustained, period of engagement with a site *and* its users *in situ.*

I did not set out to imagine a design proposal. This project allowed me to question practices of landscape architecture. In this chapter, I move toward a deeper and lighter design touch, questioning the assumption that designers always need to make a physical intervention. The practitioner sits beside the educator and the researcher. Writing text is one of the ways that landscape architects can help to shape the world. My fieldwork in Bahia was not intentionally designed, unlike other projects I worked on. It happened and unfolded and was nevertheless informed by my previous experiences. Research is often driven by intuition, which is itself informed by previous life experiences.[6] In this chapter I bring you with me on a journey through an Afro-Brazilian sacred grove and speculate on what this might mean for landscape architecture.

I first learned of the Casa de Oxóssi when I accompanied Mateus there. Mateus, an anthropologist initiated into Candomblé, is what the anthropologist Lila Abu-Lughod terms a "halfie." Abu-Lughod writes, "Because of their split selves, feminist and halfie anthropologists travel uneasily between speaking 'for' and speaking 'from.'"[7] And this "in-between" perspective was very helpful for me. For over two years on and off, I spent time in the *terreiro* with Mateus and inhabited its landscape. In his company, I was exposed to the magical aspects of these groves, and the conflicts within them, I simply wanted to learn more about the spaces. I saw this as "prefieldwork," a period of immersion, trying to understand if it was a topic I should devote more time to. This unofficial fieldwork ended not with a decision from me,

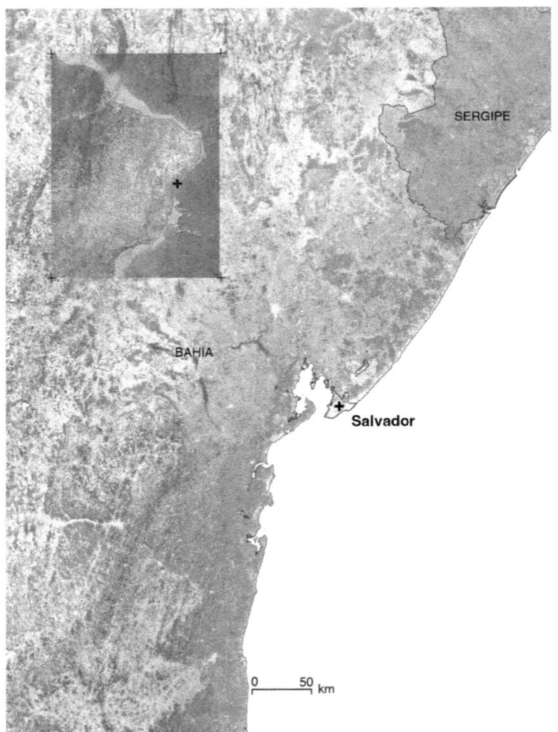

Figure 71. Salvador da Bahia, in the Brazilian northeast. (Maria Vollas)

but because of an intense experience in "the field" that convinced me to commit to studying the *terreiro* as a serious line of inquiry. This project taught me that the distinction I had made between prefieldwork and fieldwork is actually erroneous. With my fieldwork over a period from 2016 to 2019, there is not such a clear line between prefieldwork and fieldwork since there's not some distinct, bounded year dedicated to fieldwork. What I had considered prefieldwork at the time, was in fact landscape fieldwork.

Urban Green Spaces

"When I see a bit of green in Salvador, I tell my clients it must be a *terreiro,*" Alice told me as we chatted in the front yard of the Casa de Oxóssi. Alice owns an Afrocentric tour company, and the core of her business is bringing clients to *terreiros* to experience the intense ceremonies involving music, dance, and the entrancement of some participants. A crucial feature of long-established *terreiros* in Salvador is that they typically occupy a luscious green expanse in the midst of an urban setting, often around the same size as a soccer field. *Terreiros* are usually filled with plants and greenery because leaves are so important for Candomblé.[8] While some *terreiros* are greener than others, these large spaces inevitably have an impact on urban ecology and establish important relationships with the surrounding communities, which is a central concern in this chapter. *Terreiros* are among the last remaining green spaces in increasingly dense Brazilian cities, and they provide essential ecosystem services. These landscapes, however, are little understood spatially, materially, and ecologically, and rarely discussed as a particular land use within the context of landscape architecture, urban design, and city planning. Many Brazilians have never been inside an Afro-Brazilian sacred grove.

Why are *terreiros* so often overlooked? Until the late 1960s, Afro-Brazilian religious practice was illegal in Brazil.[9] Today, elements are still held in secret; it is risky to practice such African ancestral traditions within the wider national racist and anti-Black context. It is telling, from that perspective, that Afro-Brazilian sacred spaces are called *terreiros* in Portuguese, meaning yards. A yard is not valued as a garden, park, or sacred grove might be. The *terreiros* bear some equivalence with the sacred groves of West Africa where Candomblé traces many of its roots, but in

Figure 72. Terreiro Vodun Zo. (Photograph by Leonardo Finotti)

contrast to these public groves, *terreiros* function as secret parks.[10] *Terreiros* were often built in the countryside, on the city's periphery, on elevated sites, surrounded by trees to muffle the sight and sounds of the music and drums.[11] Rituals that are performed openly in Nigeria are, in Brazil, often conducted behind closed doors, away from the public gaze. The French anthropologist and sociologist Roger Bastide writes of *terreiros* as separate spaces within the city: "Ce sont des mondes à part, des espèces d'ilots africains, au milieu d'un océan de civilization occidentale, et non un continent, un bloc bien soudé" (They are worlds apart, African islands in the middle of an ocean of Western civilization, not a continent, a tightly knit block).[12] While it is fair to describe *terreiros* as worlds in their own right, they are fusions of Africa and Brazil, the outcomes of colonial processes, rather than African islands per se. And while they do function as secret parks, separated from their proximities, they are often well-knit into their ecological contexts.

Although they are often misunderstood, *terreiros* are plentiful in Brazilian cities, especially in the Northeast, and particularly so in Salvador da Bahia, one of the largest cities in Brazil.[13] Officially, there are currently about 1,155 *terreiros* in Salvador da Bahia—although some scholars suggest there are more than 2,000, a number underestimated because of systemic racism.[14] Despite this, Salvador is traditionally known for being the city of 365 churches, having a church for every day of the year.[15] Salvador de Bahia is commonly referred to as "the Black Rome" because of a combination of factors, including the number of churches, the city's hilly topography, and the fusion of Catholicism with the Yoruba religions that enslaved populations from West Africa brought to the New World. Between 3 million and 5.5 million enslaved peoples are estimated to have been forcibly taken from West Africa to Brazil, and Salvador was a center for the slave trade. In contrast, 300,000–430,000 enslaved peoples were brought from West Africa to the United States.[16] As the last country in the world to formally abolish slavery, in 1888, the memory of slavery in Brazil is still quite raw. Older generations knew people who were formerly enslaved, and the horrors and hierarchies that made the system of slavery possible are still to this day deeply ingrained in Brazilian society. The *terreiros* are one of the most visible legacies of the slave trade, providing a home for the disenfranchised and displaced, and descendants of enslaved peoples who to this day are on the margins of society.

Terreiros are spaces for other marginal groups too. *Terreiros* are considered gay-friendly landscapes and Candomblé a gay-friendly religion. Many *terreiros* have historically welcomed sexual and gender diversity.[17] A high proportion of Candomblé practitioners identify as queer, supported in no small part by discussions regarding the gender and sexual fluidity of some of the orishas themselves, such as Oxumarê, who reportedly spends part of the year as a man and part as a woman.[18] In explaining the openness to the lesbian, gay, bisexual, transgender and queer plus community (LGBTQ+), a Candomblé priest, Baba Ogu Dare, declared: "The concept of sin is a Christian concept that does not exist in the pure religions of Africa. According to the African oduns everyone is born predestined with a fate that is already set out for him, so if a human being is born with a heterosexual, bisexual or homosexual predestination, this is already in the plans of creation."[19] In addition to welcoming the LGBTQ+ community, they also support Afro-Brazilian, antiracism, and environmental movements. The anthropologist Mattijs van de Port points out in his article "Candomblé in Pink, Green and Black," that *terreiros* attract artists, intellectuals, media celebrities and politicians, not to mention many anthropologists.[20]

The anthropologists are often white and tend to study the phenomenon of trance. *Terreiros* impact various strata of Brazilian public life and touch the urban political ecologies of Salvador da Bahia in extremely complex ways.

As Salvador and similar cities expand, new developments threaten and often encroach upon the green spaces of *terreiros.* Nearby high-rises disrupt the privacy valued by these spaces, while conflicts arise from the sounds of the religious ceremonies punctuating the night. Market forces combine with institutional racism through government and elected officials, pressuring practitioners of these sacred

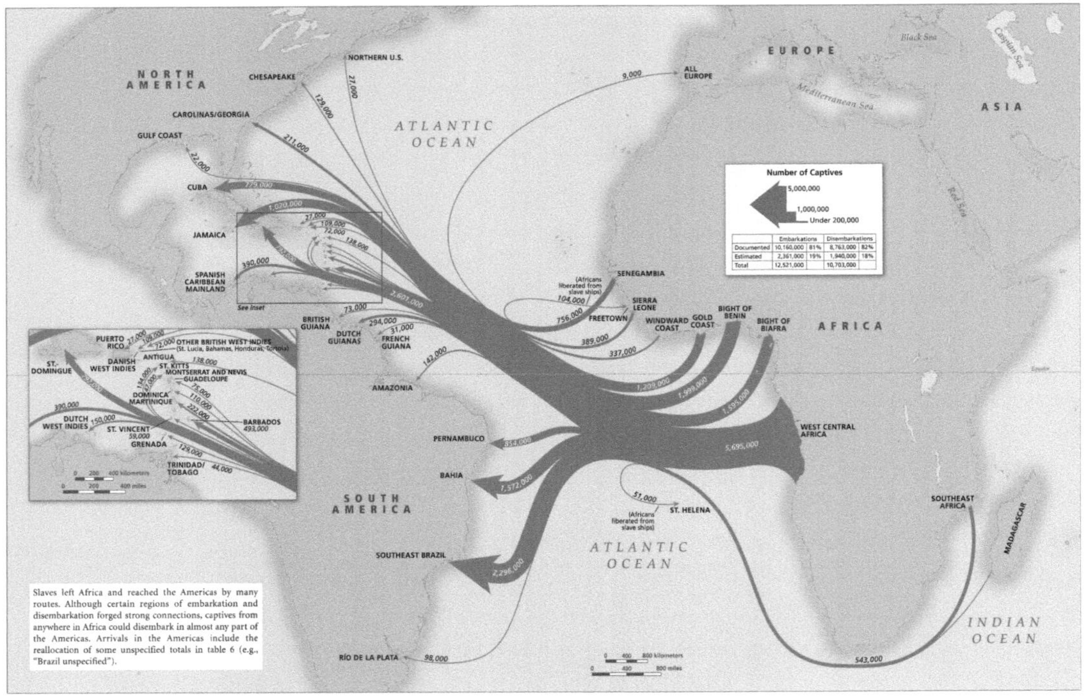

Figure 73. The slave trade out of Africa. (From Eltis and Richardson, *Atlas of the Transatlantic Slave Trade* [New Haven: Yale University Press, 2010])

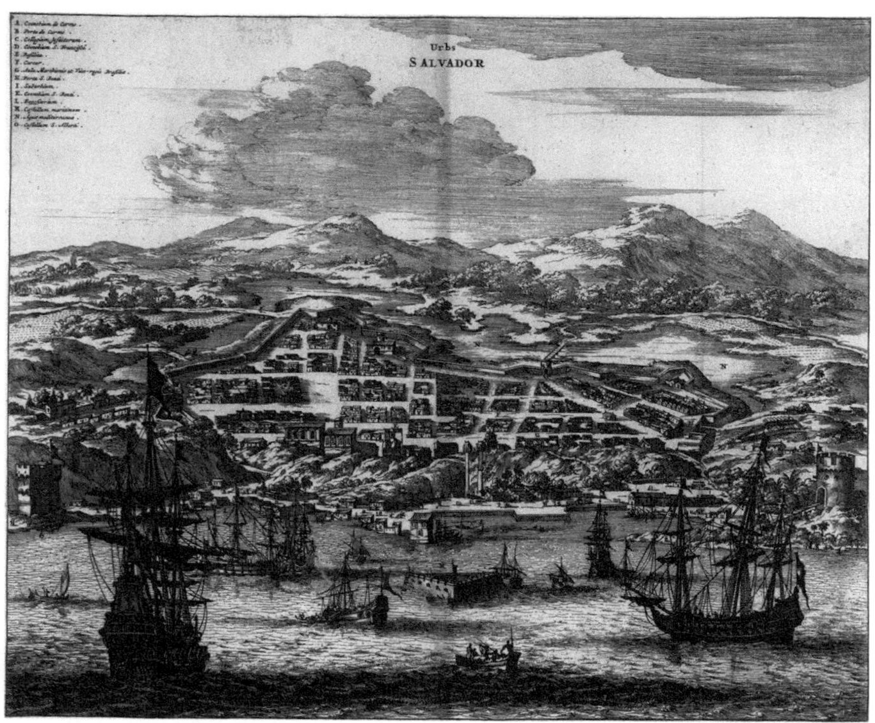

Figure 74. Salvador da Bahia, ca. 1671. (Montanus, Arnoldus, 1625?–1683. Urbs Salvador. [Amsterdam, J. Meurs, 1671] Map. https://www.loc.gov/item/2004632029/), World Digital Library)

groves to give up their green spaces. Even Afro-Brazilian rituals that are by necessity performed in the public domain are threatened by acts of religious intolerance. The tension between the sacred/secret and the profane/public was palpable during my fieldwork.

Experiential Knowledge

In this chapter, I want to emphasize the importance of the experiential knowledge that comes from an extended but discontinuous period of engagement with a particular location.[21] My fieldwork was conducted not as one preordained period in a particular site, as in the classic model of ethnographic fieldwork, but as an accumulation of shorter visits over several years. When I had two weeks to spare, I would go to Salvador da Bahia. I purposefully tried to participate in as many of the main festivals of the year as I could. Such a form of engagement is close to what Albena Yaneva terms "slow ethnography," when she studied the work of OMA (the Office for Metropolitan Architecture, based in Rotterdam) over two years.[22] Yaneva distinguishes the slow ethnographer from the hasty sightseer. The latter races through a building, or fieldsite, takes a picture to "provide her with the possibility of coming back and slowly discovering all those features that the swift moment of perception hampered her from seeing."[23] The slow ethnographer will by contrast "move about, within and without, and through repeated visits, she will let the building gradually yield itself to her in various lights, speeds and intensities, and in connection with changing moods, crowds of people and flows of things."[24] For academics it is often difficult to spend a year or more in the field since institutions can frown upon faculty taking too much time away from teaching and administrative responsibilities. Instead, academic researchers must adapt to doing what they can in shorter periods over an extended period. In professional offices, relationships with clients often extend over years, and similar long-term arrangements can be formed.[25] This slow way of doing ethnography in academia as in professional offices has myriad advantages, including more time to build relationships, and to reflect.

I recall one day early in my fieldwork—what I had erroneously considered "prefieldwork"—when I visited the Fundação Pierre Verger in Salvador to consult the photographic archive and the extensive library there. I had become fascinated by the Afro-Brazilian sacred groves, though I realized how difficult it would be to pursue this topic given all the secrecy, restrictions over access, and language barriers. At the foundation, I realized that there was a huge body of scholarship that could help me with my research. I felt an immense sense of anticipation as I learned more about the life and work of Pierre Fátúmbí Verger. Verger was a French photographer and self-taught ethnographer who did fieldwork in West Africa and Brazil. He eventually settled in Salvador, where he became both a Candomblé initiate and a professor at the Federal University of Bahia. The foundation is in Verger's former home and contains his papers and over six thousand photographs, many of which

Figure 75. Terreiro do Gantois, 2018. (Photograph by Leonardo Finotti)

document the African diaspora. With landscape fieldwork, we don't know where it will take us, or what the eventual result will be. That day I realized I wanted to study Afro-Brazilian sacred spaces, even though I did not know where it would take me.

Giving form to the amorphous is one of the great challenges of landscape fieldwork. As in other projects, I conducted my research through the embodied engagement of being in the site and actively engaging with the space. But in other ways my landscape fieldwork in Bahia was radically different to other projects I have done. It was distinct because it centered largely on one space—the Casa de Oxóssi. Although I visited other *terreiros* and occasionally archives such as the Fundação Pierre Verger, the fieldwork focused on this one confined urban space. Being immersed in that space also allowed me to sense the wider urban ecologies of the city. Before addressing the urban ecologies of *terreiros* more generally, then, let us next touch on the spatiality and materiality of the Casa de Oxóssi.

The Home of Oxóssi

The Casa de Oxóssi is one of the most distinguished *terreiros* in Brazil. It is a mother house, meaning that other Candomblé houses founded in São Paulo and other cities are seen as daughters or sons of the Casa de Oxóssi. It is among the oldest terreiros in Salvador. The *terreiro* is formally known in Brazilian Yoruba as the *Ilé Ọ̀ṣọ́ọ̀sì Àse Ṣàngó.* This can be roughly translated as the Home of the Orisha of the Forest sustained by the power of Ṣàngó—also known as the orisha Xangô in Brazil, the orisha of fire and justice. It is abbreviated to the *Ilé Àse Ọ̀ṣọ́ọ̀sì,* or the Casa de Oxóssi. Before we enter the site, it is important to reflect on the differences in meaning between the words *casa* and *ilé,* which can both be translated as "home" in English. The Casa de Oxóssi can be misunderstood to be referring to a building, in this case the main religious house, but it also refers to the family descended from the house. Thus, when someone says they are from the Casa de Oxóssi, it more accurately means they are part of the extended family of that House of Oxóssi, like the House of Orléans-Braganza, the former Brazilian imperial family. The "casa" in Casa de Oxóssi might also be understood as a landscape—but the Portuguese language, the official language in Brazil, is not very kind to the concept of landscape: there is no direct translation of "landscape" in Portuguese.[26] The Casa de Oxóssi is a landscape, consisting of people and space, rather than a building, that contains multiple structures and spaces for various purposes, spiritual and temporal, as well as trees, shrubs, and wells. In Brazil, *terreiros* are often prefixed with *ilé and àsé, such as the Ilé Àse Ọ̀ṣọ́ọ̀sì. Terreiros* are examples of landscape architecture built with measurements taken from spiritual references.

The following description of the entrances and my arrival at the *terreiro* is taken almost verbatim from my headnotes of August 2017. When I visit the Casa de Oxóssi, I do not carry a notebook, a phone, or a camera, so many of my scratch notes and fieldnotes are composed in my head as I perform repetitive tasks such as sweeping the floor or washing dishes. Typically, landscape fieldworkers should keep a recording device such as a notebook on hand to scribble scratch notes while in the field. Then they should write up their notes into a narrative, and sketch, draw, interpret at the earliest possible opportunity. My fieldwork in this case was not conducive to taking notes, or sketching, though it was punctuated with long periods of silence during which I had time to think and reflect on the space and what I was seeing, had seen, or done. It also allowed me to compose my fieldnotes in my head, which I later wrote down.

The text reproduced below was written while I was enclosed in the Casa de Oxóssi. Alone in a room for three days and two nights, I underwent a period of isolation in the *terreiro* to cultivate my personal energies. During that time, when no trappings of the outside world were allowed—I even had to leave my clothes behind and don new, white, clothes after taking a purifying herbal bath—I had a lot of time to think and write "headnotes." As I lay on the floor watching the shadows on the roof tiles (the only way I could sense the time of day or night), I became intensely aware of the sounds within the *terreiro*—from the crow of the roosters and cries of other animals to the murmurs of community members talking, to the sounds of the orishas themselves. Through the sounds, combined with my knowledge of the space prior to isolation, I began to audibly understand the structure and flow of the space.

Figure 76. Terreiro do Bogun, Salvador da Bahia, 2018. (Photograph by Leonardo Finotti)

Entering the Grove

Looking at the Casa de Oxóssi from the other side of the road, it sits as a "cloud of green" on the hillside.[27] This green cloud bursts out from behind a two-meter-high wall, surrounded by apartment buildings and a favela on the sides and to the rear. The front wall had been white until it was regularly covered in graffiti and messages such as *Jesus é o caminho,* "Jesus is the way." Such acts of religious intolerance are not uncommon toward this community, and Candomblé communities more generally in Salvador. The community's response was to commission their own graffiti to cover the bigotry, so now they have a graffiti-covered wall rather than the white wall they had intended. White is very important for Candomblé and permeates the city. One may consider the white taxicabs of Salvador a public display of the energy of the orishas. On Fridays, when it is customary for followers to wear only white, as Friday is Oxalá's day, the streets and shopping malls of Salvador are strikingly monochromatic. Oxalá is the most senior of the orishas and his color is white. In extreme cases, people wearing white become targets of intolerance; and I have been subject to such abuse simply because I was wearing white Candomblé clothing in a public space.

There are two primary entrances to the Casa de Oxóssi. The most impressive is the flight of steps leading up from the narrow blue gate on a main thoroughfare that both connects and divides Salvador.[28] Entering the blue door from the street—Oxóssi's color is light blue—the shrine to Oxum, the orisha of love and sweet waters, greets the visitor immediately to the right. This shrine has a well that is connected to a river system that formerly flowed outside, before it was undergrounded and largely forgotten during the development of the city. A road now runs directly above the undergrounded river. Salvador's various rivers would originally have been partly responsible for the locations of the *terreiros.* Rivers, as sources of sweet water, are associated with Oxum. I wonder if Oxum, who is beautiful and likes to be admired, is sad that the river is today hidden beneath the roaring highway.

The generous steps from the outside road to the Casa de Oxóssi are punctuated by seven small landings. As one climbs the hill, on each side one finds small structures and gardens devoted to individual energies: the energy of time, the energy of trees, and the energy of chance. Most, but not all, of the houses are on the right-hand side as one ascends the flight of steps, and much of the left-hand side is devoted to guest houses, offices and a yard for livestock and poultry, mostly goats, chickens, and sometimes tortoises, pheasants, and peacocks. The livestock is not raised there but purchased in response to specific ritualistic needs and as food for the energies and for the community. On reaching the top step, one arrives at the yard in front of the main building.

Arrival Rituals

I usually approach the *terreiro* from the other gate, which is level with the outside street, and leads straight to the front yard. When I arrive at the *terreiro,* I leave behind the world outside. Using a half gourd, I draw water from a large earthen container outside the gate on the street. Three splashes of water from left to right purify my entrance, first at 10 o'clock, then at 12 o'clock, and then 2 o'clock. Next, I knock on the gate, not expecting a human answer, but to alert the energies of my arrival. I walk, head down, to the rear of the house to take a herbal bath. Until the bath, I am effectively invisible to the community. I refrain from greeting or making eye contact with the people I surely meet scurrying around. No one talks to me. Well, no one should talk, and if they do, I should not reply to them. Sometimes, though, members will exchange a secret smile with me: it's hard not to reciprocate when someone smiles.

The bath that I enter contains various concoctions of leaves, such as those recorded by Pierre Verger in his monumental book, *Ewé, the Use of Plants in Yoruba Society.*[29] Verger meticulously documented sacred leaves and the possible outcomes their combinations produce.[30] The bath, which can have quite an intense pungent smell, not only washes off the negative energies of the outside world and purifies the body, but it gives *axé* to the bather. *Axé* (from *àṣẹ,* in Yoruba) is the energy from which all life flows—an essential concept at the core of Candomblé.[31]

Once I have bathed, I do not use a towel to dry off the bathwater. Rather, I must wait for my wet body to dry itself, which doesn't take long in the Salvador heat. Once I'm dry, I change into clean white Candomblé clothes. The clothes are white for everyday purposes. During a festival, initiates will wear the color of their orisha; everyone has a primary orisha who "owns the head." The head is the most important body part according to Candomblé, and where the orisha or energies dwell. I am

Figure 77. Orisha house, Terreiro Vodun Zo, Salvador da Bahia, 2018. (Photograph by Leonardo Finotti)

not allowed to wear shoes. Contact with the ground is an essential aspect of Candomblé. For the lowest in the hierarchy, constant contact with the earth is a must. The *axé* of the house is buried in the ground; it's where the dead dwell. The reddish soil, rich in iron oxides, stains my feet red. The concrete that covers much of the working spaces in the *terreiro* is painted red, and the earth doesn't stain it as much as it would have if it were grey.

Once purified, clothed and barefoot, I am ready to greet the energies. The orishas are housed in different locations in the *terreiro,* where they are fed and cultivated through various rituals often conducted in secret. It is my duty to greet them with a bow to the floor and the forehead touching the ground. At the Casa de Oxóssi, I first must greet Ogun (the orisha of metals and war), followed by Oxóssi (the orisha of the forest, hunting, and art) and Oxum (orisha of sweet water and maternity), who are housed together in a single structure. Next, I greet Obaluaye (the orisha of earth and contagious disease) followed by Yemanjá (orisha of the sea, syncretized with the Virgin Mary) and Xangô (the orisha of fire and justice). I greet Exu (the orisha of chance, movement, and communication), Iroko (the orisha of time), and Ossain (the orisha of leaves and concoctions) from the main yard. Next, I enter the main religious house. First, three backward sweeps of my feet in deference (again) to Exu, the orisha of chance. Then, I prostrate myself before Oxumarê, the orisha of the rainbow and the snake, and Iansã (the orisha of the wind). This is followed by greeting Oxalá, the orisha of the air and the father of all the energies. Oxalá is my orisha, the owner of my head, and I prostrate myself completely on the floor when greeting a special energy such as Oxalá. Finally, I greet the *axé* of the *terreiro,* around which the initiates dance, then the ancestors, and then the drums. Crucially, the drums make the energy of the house audible and propitiate, or provoke, trance. The sequence of the greeting and the spatial arrangement of the orishas is very deliberately determined. Exu, for instance, is a trickster who is always celebrated first in the ceremonies, since if he is appeased from the beginning, he is less likely to interfere in the proceedings. This might be why he is greeted twice. It is important to consider that even though the orishas are greeted in this way, they are not necessarily located beside one another. This requires a certain back and forth on behalf of the greeter. Ogun, for example, is housed in between Yemanja and Xangô. In Brazil, there are sixteen basic orishas, plus one.[32] This is much simpler than the West African pantheon of 201 or 401 gods.[33]

Then, I greet the human community, beginning with the leadership. Iya, the *Iyalorixá* (*Ìyálòrìṣa* in Yoruba) mother of the house, or *Mãe de Santo* (Mother of Saints), comes first. Candomblé leaders are often women, which is one of the reasons Ruth Landes's insightful book on Salvador da Bahia is entitled *City of Women.*[34] It is important to see and be seen in the *terreiro* and, by the time I have greeted the energies, I usually have a sense of whether Iya is in residence. Iya's presence creates a flurry of activity, and her absence a lull. If I am unsure if Iya is in the *terreiro,* a good place to start looking for her are her rooms at the corner of the main house that frame the central yard. Like one of the orishas' houses from the outside, it is like a lifeguard's post with an eye to the yard. It is easier to see out of the rooms than to see in toward the darkened space from the outside. After greeting Iya, I find each member, starting with the second in ranking and going down the ladder. I bow before each, arms outstretched, and hands cupped, asking for a blessing, and say in Brazilian Yoruba, "*Motumbá.*" The answer, invariably, is "*Motumbá axé.*" I am told that at the Casa de Oxóssi, *motumbá* is said to mean something like, "I submit myself to you!" and the reply is "You are submitted to the power of the axé!"[35] *Motumbá* appears to be related to *mo júbà* in contemporary Yoruba in Nigeria, meaning, "I pay homage." The elders may give me a hand to kiss or offer a hug. Hugs are very important in Candomblé, they embody the transfer of energy, *axé.*

After greeting the community, I begin to work for the maintenance of the house

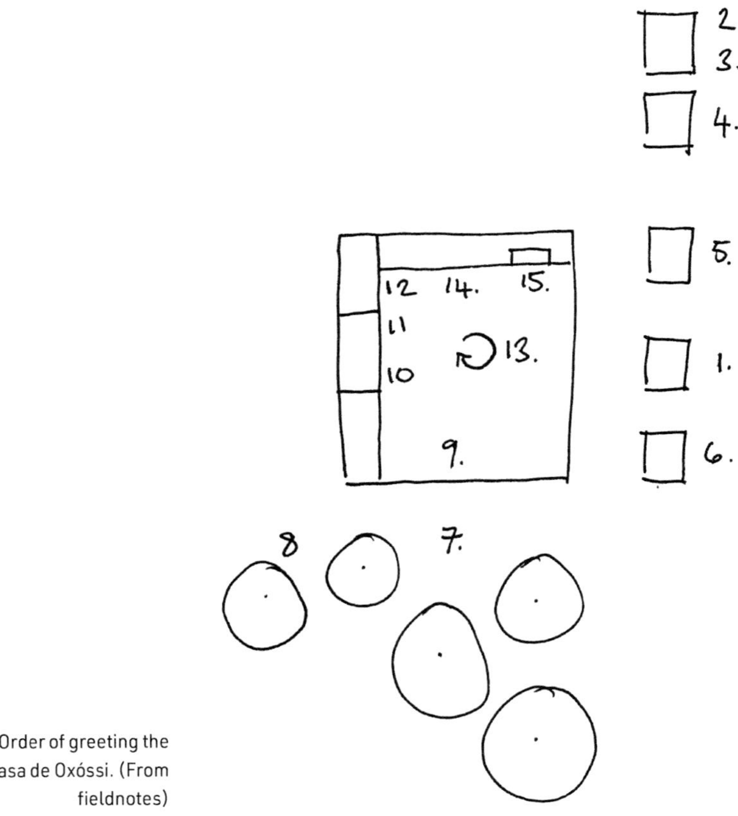

Figure 78. Order of greeting the orishas, Casa de Oxóssi. (From fieldnotes)

or in preparation for the rituals. The rituals may either be closed to the community or performed as public ceremonies. It is the former, the secret rituals of the houses, that are, paradoxically, increasingly performed in public spaces due to the spatial and vegetal limitations of the *terreiros.* The public ceremonies are usually scheduled for 8 in the evening, after the sun goes down, although the rituals generally do not start on time. I hear many people say that Candomblé has its own time. It is telling that there are no clocks or mirrors in the *terreiros,* and it is easy to lose track of time and even of oneself as one gets immersed in the life of the community. There are many tasks to be done, whether in the kitchen, preparing food for the orishas, or, as in my case, washing the dishes and sweeping the floor. I find sweeping the floor to be particularly helpful for getting to know the space. Sweeping my way outside from the central space of the *terreiro,* I have become acquainted with the tiles and cracks on the floor. But sweeping also exposes me to the community, and I meet people through the act of working. As I sweep my way from the inside of the *terreiro* outward to its periphery, I get to study the materiality of the space.[36]

Gathering Leaves

Other important tasks involve gathering leaves or combining leaves to prepare baths. When preparing for the festival of Oxalá, I am often tasked with going outside the *terreiro* to help harvest leaves, gathering fronds from young palms beside the highway and thus avoiding the need to climb the taller palms inside the *terreiro.* Considering the sheer diversity of plants needed for ceremonies and rituals, most *terreiros* harvest the leaves from outside. "There are leaves that need to be picked (outside of your own space)," says Vilson Caetano de Sousa Júnior, an anthropologist and a *babaorisha* or "father of saints."[37] Plants, or leaves (the terms leaves and plants are often used interchangeably), are important sources of *axé;* they even have their own dedicated religious position within the community. The *babalossain,* the "father of leaves," takes care of the leaves. Ossain is the orisha or energy of the leaves.

Sousa Júnior confirms the essentiality of plants in Afro-Brazilian religions when he writes: "Não existe candomblé sem folha," meaning, "There is no Candomblé without leaves." Sousa Júnior also repeats a Yoruba saying, "Kosi ewé, kosi orìṣa":

Figure 79. Gathering leaves along a highway. (Photograph by Leonardo Finotti)

"Without leaves there is no orisha."[38] Many of the specific plants associated with orishas are not the same in West Africa and Brazil; enslaved people found equivalent species in the New World when the species formerly associated with the orishas were not available.[39] From fieldwork, I gather that in Nigeria the *iroko* tree (the embodiment of the orisha of time) is African teak (*Milicia excelsa*); in Brazil the tree that has this function, and which is called "iroko" there, is a fig, *Ficus insipida.* Sousa Júnior talks about the ritual process called *sasanhe,* or "singing leaves," where the leaves are themselves enchanted through complex rituals that take place at dawn, an important time in Candomblé when the positive forces of the new day beckon. According to Sousa Júnior, the *axé* of the leaves needs awakening before it is imparted.[40] Another important reference is the book *O que as folhas cantam (para quem canta folha)*—"What the leaves sing (for those who sing leaves)"—by Mãe Stella de Oxóssi, one of the most famous Candomblé leaders in recent memory.[41] In Candomblé, leaves are classified as male or female, and as either hot or cold.

Figure 80. *Milicia excelsa* (Welw.).
(Courtesy, Meise Botanic Garden)

Figure 81. *Ficus insipida Willdenow.* (Maria Werneck de Castro, courtesy National Library of Brazil)

Male leaves are long, while female leaves are rounded. Hot leaves awaken, while cold leaves have a calming effect. A son or daughter of a cold orisha cannot use the leaves of a hot orisha and vice versa. Nettles are an example of hot leaves, while cold leaves include basil and lavender. Some leaves, such as the arrowleaf elephant's ear, are both male and female.[42]

Figure 82. *Alocasia lowii var. picta*, elephant ear. (Curtis's botanical magazine and Biodiversity Heritage Library, Public Domain)

The Turning

One encounter during my prefieldwork in the Casa de Oxóssi stood out as a turning point for my landscape fieldwork. Every year in June, the Casa de Oxóssi celebrates the festival of Oxóssi. The community spends weeks in preparation for this splendid event, or, more accurately, sequence of events. I helped with the preparations, which included replacing all the crepe paper on the ceiling of the religious space, and attended most if not all the preparatory rituals. The celebrations culminated in the festival of Oxóssi on the Saturday evening, beginning after dark at 8 or 9 pm and extending until the early morning. It seemed like everyone from the community was there, including the elders, and many Paulistas, as people from São Paulo are called, sons and daughters of the House who flew in for this series of festivals and rituals in the days before and after the main event. (It was easy to distinguish the Paulistas by their airs and graces and more elaborate clothing, which Soteropolitanos—people from Salvador da Bahia—often find to be of bad taste.) The ceremony is followed with a huge breakfast served outside well after midnight.

The public ceremony was followed by a ritual the next morning, restricted to the community. This ceremony, which involved the killing and sacrifice of goats and chickens, started mid-morning, and as is typical for Candomblé ceremonies went on for hours. Despite the intensity and the beauty of the music, I started to get very restless. I was sleep-deprived after attending the ceremonies and participating in the preparations for the ceremonies over the previous few days, often late into the night. I had also been sleeping poorly on a straw mat on the floor in the *terreiro.* Maintaining contact with the energy of the earth is not always very comfortable. I also did not understand the structure of the ceremonies. At that stage, no one had explained to me that they begin with appeasing Exu, followed by invoking the other orishas through dancing around the *axé* of the house, before the orishas begin to descend through trance. I was also deeply disturbed by the animal sacrifice. I stood in the public gallery beside three elegant goats who were wearing ribbons and dresses and accompanied by many chickens. First, Iya asked the animals one by one for their permission to be sacrificed by offering them leaves to eat. Should they accept the leaves, this was taken as a sign of acceptance of sacrifice, and they each acquiesced and were taken off for their throats to be slit by the *ogans. Ogans* are male members of the community who do not enter trance and who are initiated to Oxalá, the orisha of peace and wisdom. Thankfully, the sacrifice took part in an adjacent room. Even that I felt was too violent for me. I reminded myself that animals are killed all the time for food, but still struggled to accept it. What did ritual sacrifice have to do with landscape architecture anyway? All I wanted was to go home and rest.

Then, I noticed a change in the intensity of the music. It seemed to be slowing down. Maybe the lengthy ceremony was ending? As the intense beat from the drums abated, Iya, the mother of saints, emerged from a side door. Iya was dancing. As she got closer, I could tell that she was entranced because her eyes were shut, and she was wearing a white cloth tied around her waist. I knew that this was the cloth that the *ekedies* place on the individual when the orisha descends.[43] *Ekedies* are the female equivalent of the *ogans,* women who don't enter trance and who take care of the orishas. I had noticed this procedure in the ceremonies and associated it with the arrival of trance. Another sure sign of entrancement was that Iya's arms were folded behind her back, in the usual pose of the orishas. Iya had clearly incorporated the energy of the forest. At this stage, I knew that Oxóssi was Iya's orisha—perhaps one of the reasons her destiny as *Iyalorixá* was determined from an early age. I was told later that this was a special moment in the life of the house. Senior members of the community do not get entranced very often, and Iya only gets entranced once or twice a year. Iya had gone away and in Iya's place dancing on the floor was Oxóssi. (I will now refer to Iya, who is embodying the energy of Oxóssi, as he.) He was wearing white clothes, and with arms and shoulders uncovered, and he was sweating profusely. A white turban was wrapped around his head with splashes of red on the front, freshly stained with the blood from the goats and chickens that had just been ritually sacrificed. Fresh blood is one of the most potent sources of *axé*. The music geared up again, and in between intense dances back and forth across the main religious space, Oxóssi hugged those close to him. In due course, Oxóssi danced his way to the public gallery where I was standing. Although the ceremony was only open to affiliates of the house, there were several others removed from the main gathering, guests and uninitiates like me. We were strangers; I had never seen them before. Oxóssi hugged each of them, but Oxóssi did not hug me. Why? Why did Oxóssi not hug me? Did my bad energy repel him? Did he sense my boredom? Maybe he didn't see me, as his eyes were closed? I felt disappointed. I had been standing here for hours. I had helped with the preparations, for weeks, yet Oxóssi had hugged these people I had never seen before instead of me!

The rhythm of the music slowed down again, and a queue formed in front of Oxóssi, who was now near the front of the room. The unhugged were getting in line to receive hugs. Clearly, Oxóssi had missed hugging others too. I didn't feel quite so neglected, but still I wanted a hug. I watched as Oxóssi embraced the people in the line one by one: first to the left, then to the right. I asked Fábio, one of the initiates, if as an outsider I could join the queue. "*Certo!*" Fábio replied, gesturing for me to join the queue. I wasn't sure whether to keep my flip-flops on or not—as a visitor I could wear shoes. I think I kept them on. As soon as I entered the line, Oxóssi came straight toward me—eyes shut—passing by the ten or fifteen people in front of me. The orishas move with their eyes closed and yet they navigate the leaf-covered floor seamlessly.

Oxóssi embraced me firmly, his head to my right. He hugged me again, with his head to my left. The second embrace seemed to take a very long time. Then, Oxóssi let go and touched his forehead on mine. The fresh red bloodstains, filled with *axé,* were pressed against my forehead. Once. Twice. Three times. On the third touch, I could not move. There, I was standing right in the middle of the room with my limbs frozen. Numb. Transfixed. *Totalamente congelado* was how I later described the moment in Portuguese. It doesn't translate so well in English: "Totally frozen." My limbs were frozen, but my body felt warm. Afterward, people told me that after touching my forehead Oxóssi had also delicately touched his right hand on my left cheek before he danced off to give more hugs. I am told that when the community saw what had happened, they immediately started looking for Mateus to take care of me, shouting, "*Seu amigo virou!*" This meant, "Your friend has turned!"

When I came around, sometime later, I was lying on the floor. They had lain me on top of a straw mat to recover, as is customary, keeping close contact with the ground. As I opened my eyes, and my surroundings came into focus, I saw two pairs of eyes looking down at me. Mateus had brought Pai João, one of the leaders of the house. I recall Pai's beady eyes staring at me intensely for quite a while as I lay motionless on the floor. Pai declared, "*Energia boa! Energia boa!*" meaning "Good energy! Good energy!" He mumbled some instructions to Mateus and walked away. I recall Mateus saying, "This is not happening. Please, tell me this is not happening. This is not supposed to happen!" before bringing me water to drink. For Mateus, I was a curious outsider, a white foreigner, a "gringo," and the turning was clearly surprising for him. I was not surprised because I still didn't know what had happened. Meanwhile, others came to look at me lying on the mat. Among them was the senior *ogan,* who inspected me and said he sensed the work of Oxalá, my orisha.[44] At that point, I did not understand what had just happened to me. But in the eyes of the community, I had been very publicly touched by the energy of the orishas.

I spent the rest of that afternoon and evening at the *terreiro* not comprehending the event, or even knowing if there was much of an event. I knew that something had happened to me, but I was not sure what. I had overheard someone saying, "É a primeira vez!," "It's his first time." But I was not sure if they were speaking about me, or what was first in any case. Nobody explained anything to me, maybe because it is not something that is easily spoken about, or because it is just normal in the

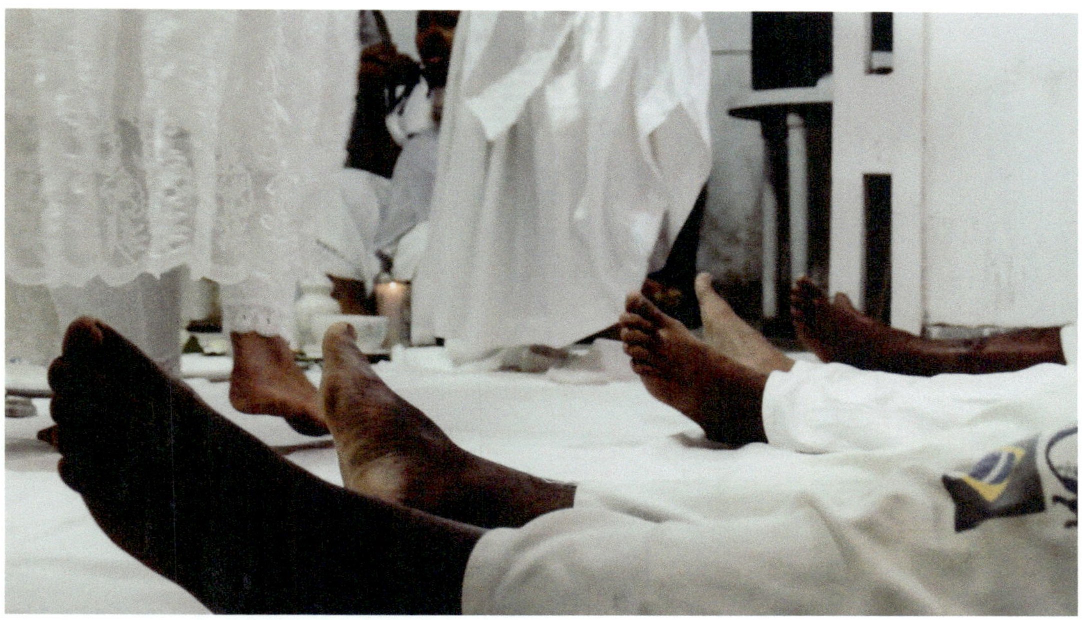

Figure 83. "Candomblé: Oblation to the Gods." (Courtesy of Sam Campbell)

terreiro. Mateus was busy and assigned to working in the garden, away from me. That evening, while still in the *terreiro,* I realized something had changed within me. It was as if my brain had been taken out of my head and washed under a shower and then returned, retuned, and refreshed. It was not "brainwashing," because I had no doctrine or set of beliefs to be brainwashed with. But my mind was filled with a positive energy displacing my doubts and negativities and worries at that time. When I realized this, I remembered what Pai João had said: "Energia boa! Energia boa!"

Committing to Fieldwork

It was to take some time for me to comprehend the significance of what had happened to me that June day in the *terreiro.* This "turning" meant that I was a *rodante,* someone who has the gift of trance. The orishas had chosen my body to dwell in. This was an important moment in the eyes of the community. After this event, I was treated differently, not necessarily any better or worse, but like I belonged. I was not the only foreigner that this happened to, but it was still an unusual happening. On hearing of my gift, one elder remarked, "Os orishas gostam des persoas brancas tambem!," meaning, "The orishas like white people too."

The day following the "turning," I returned to the United States. When I next returned to the Casa de Oxóssi a few months later, I was made an *abiã* (*abiyan* in Brazilian Yoruba). An abiã is a novice, what Mãe Stella de Oxóssi terms "a pre-initiate grade. Literally, 'one who has kin by affinity.'"[45] No one asked me if I wanted to be an abiã. After a consultation with Iya, I learned of my designation from Mateus with the words, "Congratulations, your life has just become a lot more complicated." Abiãs, Mateus explained, have a special place in the community: they are junior members, who are at the bottom of the hierarchy of the house. They participate in the life of the community and are considered members of the house. I had gone from being a curious visitor, who, on account of being a foreigner, was sometimes feted at the top table with Iya and the elders and served meals on bespoke gold-rimmed bone china plates, to an abiã. Abiãs are not allowed to eat indoors, never mind eat at the high table. When they do eat it is from an enamel plate using only their hands—no knives or forks permitted—while sitting outside on the ground. Abiãs are not supposed to sit on chairs because their heads must be lower than anyone from the community who is spiritually older or higher ranking than them—in other words, lower than everyone apart from the other abiãs and visitors.

Abiãs learn through working and doing. There are many tasks to be performed in the *terreiro,* like sweeping the floor, washing dishes, and taking care of other menial tasks, and abiãs are important labor source that keeps the *terreiro* functioning. I had volunteered to do these tasks while a visitor, but as an abiã it was expected of me to work. I realized that what I was doing by completing menial tasks was ideal for participant observation research. Anthropologists have often taken up jobs or other roles in the community they are studying to get to know a community while also giving back in some way. So, in a form of participant-observation they are actively engaging in a form of labor, care, and knowledge exchange. Such practices are connected with what the Brazilian anthropologist Marcio Goldman writes about the importance of *fuxicos,* which literally translates as "gossip," as a form of learning.[46] *Fuxicos,* he explains, "are actions meant to produce effects, a search for a pure efficacy, allowing to raise the hypothesis that cosmological, mythological and classificatory systems are all actually in service of this operational process."[47] Goldman is describing learning by assimilation, by observing, by doing, rather than by asking questions. In other words, Goldman means more than gossip when he uses the term, it is about a way of learning. While the *terreiros* are rife with gossip—which can be and is an important form of knowledge transfer for anthropologists as much as for people of Candomblé—the abiã learns through what is essentially a form of participant observation. This form of participant observation is especially useful for someone like me whose Portuguese is not fluent. Language is not as important as it would otherwise be because even with the most fluent Portuguese, the abiã should not ask questions or expect to receive knowledge directly imparted by a great master. But they listen to what they hear, and they work. Goldman also writes of the Candomblé novice: "They must patiently put together details gleaned here and there over the years, in the hope that, at some point, this accumulation of knowledge will acquire enough density to be useful. This is called *'catar folhas'* ('gathering leaves'), an expression related to the fact that knowledge and learning are located under the sign of the orishas Ossain, the master of herbs, and Oshossi, the hunter."[48] The abiã learns though a process of "gathering leaves," literally and metaphorically.

Ultimately, what was perhaps most important for me about this event, "the turning," was the insight I gained about trance. I had been fascinated but puzzled by what it meant: Was it spirit possession? Was it really an entity coming from the outside and entering the body? Or was it an inner energy resonating with the energy of the drums, the leaves, and the house? "Ecstatic encounters," is how the Dutch anthropologist Matthijs van der Port describes the phenomenon.[49] Whatever it was, as I came to experience, it was palpable, and visceral. Trying to explain trance to someone who has not experienced it is perhaps like explaining color to someone who cannot see. In *My Name Is Red,* by Orhan Pamuk, red, the color, describes color to a blind person: "Colour is the touch of the eye, music to the deaf, a word out of the darkness. Because I've listened to the souls whispering—like the susurrus of the wind—from book to book and object to object for tens of thousands of years, allow me to say that my touch resembles the touch of angels."[50] I cannot describe trance in such poetic terms. In fact, I don't have any words to describe trance other than how I described it in Portuguese: *totalmente congelado,* "totally frozen." And that is the point that I would like to make: Trance is in fact another state of being that can only be understood by direct experience. The act of the entrancement allowed me to move beyond my initial questions in a way that far transcended any reading I could have done on the topic. I could move on. At the same time, I became even more fascinated by trance as a landscape architect because of the material and spatial dimensions involved. It provided the key which allowed me to enter the space of the *terreiro.*

The Colors of the Rainbow

While I was in the *terreiro,* I often struggled with the question of how much I should actively intervene as a landscape architect or share my opinions with the community. These questions came to the fore during a time when I was helping to prepare for a ceremony. Like many Afro-Brazilian *terreiros,* the ceiling of the main religious space of the Casa de Oxóssi is decorated with garlands made of rectangles of crepe paper suspended like multitudes of small flags from taut, closely spaced wires stretched down the length of the room. Each flag is about ten by thirty centimeters (four by twelve inches) in size. Normally these garlands are all white, but the garlands of the Casa de Oxóssi's ceiling include colors, in honor of Oxumarê, the orisha of the rainbow and the snake. The colored paper flags are arranged along the wires to form a great rainbow arc that stretches from one end of the ceiling to the other. "Rainbow" translates as *arco-íris* in Portuguese. During ceremonies, these papers flutter in the breeze caused by the orishas dancing beneath. Some say that the paper garlands recall the grasses of West Africa and the orisha of the wind, Iansã.

At the Casa de Oxóssi, the paper decorations are replaced every year in advance of Oxóssi's's festival on the second Thursday in June. This is the main festival of the year, and the time for the annual thorough cleaning and tidying of the *terreiro.* One year, the group had already begun the process of mounting the decorations when I was added to the team. At that point, it was getting close to the festival and the ceiling needed to be finished. The Casa de Oxóssi's rainbow intrigued me because the spectrum of colors is not consistent with the scientific spectrum, which goes from red, orange, yellow, green, blue, to indigo, and violet. The Casa de Oxóssi's spectrum uses slightly different colors and follows a different sequence: red, orange, yellow, green, violet, blue, and pink. I wondered if this was a mistake, an improvisation, or perhaps a form of traditional knowledge handed down over generations, maybe carried across the Atlantic by the enslaved?

My assigned role was to add the glue to the top of each long rectangle of crepe paper. These pieces of paper were then handed to someone standing on a ladder, who folded the glued paper over the wires that crossed under the ceiling. It took some practice to know the exact amount of glue to put—too much and it got quite messy, too little and the papers wouldn't stick. But the bigger quandary I faced was one that many fieldworkers experience: Should I say something about the rainbow being inconsistent with the spectrum I knew? This led to broader existential questions: How much does one intervene in a landscape? Who gives the fieldworker the right to meddle in the first place? Why should my Western-derived scientific understanding of color take precedence over the community of the Casa de Oxóssi's

Figure 84. Interior of Terreiro do Bogun, 2018. (Photograph by Leonardo Finotti)

perception of the spectrum? As I mastered the amount of glue to put on the crepe paper, I wondered, what is my role in this space? Am I an observer? Or would the community welcome my voice in this symbolic manifestation of the energy of the rainbow, the community's image to the world? Confronted with this dilemma of whether I should inquire about this possible aberration, I decided to say nothing and to keep quiet and do the job that was assigned to me. I worked quickly and methodically, and it still took many hours for that purple to be added to the rainbow. Afterward, it looked good, but the spectrum still looked odd to me.

The following year, I was assigned to the same team as they embarked on the annual task of renewing the paper rectangles in the spectrum. By then, I was an abiã, a novice, which made me feel I was a bit more entitled to speak up, although I was also aware that abiãs in general should not express an opinion. This time, one of the team, a recent initiate, noticed the aberration. She had brought her mobile phone (something I wasn't allowed) and had searched for images of rainbows on her phone. When she raised the order of the colors, I felt empowered to add something too. We compared the scientific spectrum with the *terreiro's* spectrum, noting that purple does not sit beside green. But over two hundred of the papers had already been hung. No one wanted to take them down and start all over again. So, we collectively agreed, "Let's wait till next year." Next year came, and the same thing happened. I was assigned to help the group as part of my work duty. I had mentioned the story to Mateus, who immediately raised the issue with the leadership of the community by text message. Iya's approval was required to make any change to the rainbow, but Iya was traveling and unavailable. No one wanted to make the decision without Iya's blessing. So, the rainbow with the unusual spectrum persisted.

I think I did the right thing in balancing the desire to make change with respect for tradition and so-called indigenous knowledge. In some ways it doesn't matter if this was indigenous knowledge, or a simple mistake or improvisation. Only when another member of the community said something did I feel empowered to act. I knew that colors in general were central to understanding Candomblé, but I didn't know if the order of the colors mattered. While the scientific spectrum is important for its chromatic value, colors are also central to the understanding of Candomblé. Each of the orishas have their own color or preferred colors: red for Xango, blue for

Iemenja, and the white for Oxalá, which predominates, because Oxalá is the wisest and oldest of the orishas.

Trans-scapes

The act of gathering leaves embeds the Candomblé novices and initiates in the landscape. They become active participants in the landscape, shapers of it, inhabitants, and maintainers. Marcio Goldman defines Candomblé as a "set of practices and a way of life": "Candomblé should then be seen as more than a purely intellectual or cognitive system—or even as a mystical, emotional and ritual one, at least in the formal sense of these terms. This means taking seriously what every researcher into Candomblé invariably hears, namely that the information passed on to them and the festivals they freely attend are no more than the visible side of something much deeper."[51] Indeed, this concept of a set of practices can help to position *terreiros* as landscapes, if not landscape architecture. That is, as Tim Ingold writes, in the light of such practices "landscape is constituted as an enduring record of—and testimony to—the lives and works of past generations who have dwelt within it, and in so doing, have left there something of themselves."[52] In the spaces of the *terreiros,* Afro-diasporic memories, knowledge and deities are celebrated through the various rituals, including the concoctions of leaves, animal sacrifice, music, and dance. The *terreiros* are repositories of memories of slavery and places of proclamation of Black identity. Ingold tells us that to perceive the landscape is "to carry out an act of remembrance and remembering is not so much a matter of calling up an internal image, stored in the mind, as of engaging perceptually with an environment that is itself pregnant with the past."[53] The landscape is lived through the tasks that are performed there. Ingold elaborates, "Just as the landscape is an array of related features, so—by analogy—the taskscape is an array of related activities."[54] Gathering leaves, literally and metaphorically, means that you never quite know when you are finished with the accumulation of knowledge. The elders decide when they give you leaves, and when you have had enough. This is a landscape constructed not just by inhabitation and tasks but with the spiritual too. They are landscapes designed for the orishas first and people second. They are transcendent spaces. I think of *terreiros* as landscapes in which trance and transcendence, gender and materiality collide in these Afro-Brazilian islands. They are transcendent and "trance-scendent" spaces, or "*trans-scapes.*"

As I sweep the floor, beginning inside and moving my way outside, by the time I reach the perimeter it is time to start sweeping inside again—so it is a constant cycle of work. As I work, I think. I imagine how the landscape could be different. How a ceremony could be improved with a different spatial configuration. How the space of the *terreiro* itself could be more efficiently constructed—thereby exposing my own ethnocentric ideas over efficiency. One of the leaders of the house pulled me aside one day and told me she wanted to engage my professional expertise for the benefit of the community. And she had an idea. She wanted me to design an interpretative path around the *terreiro.* The path would wind through the woods and have signs explaining the spiritual significances of the vegetation to tourists. I politely replied that this could be one approach, but that this landscape architect first recommends stabilizing the slopes against landslides that are common during the rainy season; in thickening the planting of the *terreiro* to aid with soil stabilization so that more Candomblé plants are nurtured in the *terreiro.* I explained that this landscape architect was interested in understanding more deeply how *terreiros* function for both nonhuman and human audiences. We are also interested in their wider urban ecologies and how *terreiros* relate to the city at large. Since this conversation, I assembled a small research team to measure temperatures inside the *terreiros* and unsurprisingly finding the temperatures are lower than outside.[55] Lower temperatures are just one of the myriad advantages these sacred groves offer to the city of Salvador. Gary R. Hilderbrand in writing about the urban pleasures of shade, points out that urban canopy cover results in "increased rainfall uptake and improvements to storm water management, absorption of pollutants, oxygen production, carbon sequestration, and cooler summer temperatures."[56] So, while the interpretive path rightly got set aside, one hopes that by understanding the wider meaning of terreiros can help in conserving these spaces.

Meanwhile, as I sweep, I also chat with the other abiãs, with Alice, and with other community members. I get to know about them and their lives outside the *terreiro.* Even though I been focused primarily on the Casa de Oxóssi, I realized it was important to situate the site within the city to understand its wider ecologies and relationships with the city.

Typologies of *Terreiros*

The architectural historian of Candomblé, Fábio Macedo Velame, traces the internal layout of terreiros to two main sources. The Africanist approach looked to West Africa for its references from the nineteenth century to about 1980. Since 1980, Velame writes, a more cosmopolitan, Creolist approach has dominated, which embraces a wider series of referents.[57] In terms of the relationship of terreiros to the city, the Casa de Oxóssi represents a typology of *terreiro* which used to sit on the urban periphery but has more recently been subsumed by the city. It is what Bastide termed an "island," albeit an island of green, rather than an island of Africa per se.[58] The Casa de Oxóssi shares several similarities with the Casa Branca, another *island of green* that has a similar topography and history. The Casa Branca, formally the *Ilê Axé Iyá Nassô Oká,* is in fact the oldest *terreiro* in Salvador and was the first to be listed among the national patrimony managed by IPHAN, the Brazilian National Institute of Historic and Artistic Heritage.[59] This listing of the Casa Branca was not, however, an easy process as there was no precedent for including Afro-Brazilian spaces among the national patrimony listings. The architecture of the sites is considered by many as unremarkable, a reflection of limited material resources of their communities, and their importance is primarily cultural, social, and, most controversially, religious. The difficulties faced in getting the Casa Branca listed might be attributed to institutional bias and/or thinly veiled racism (it's important, in this context, to recall that Afro-Brazilian spaces were illegal in Brazil until the late 1960s). The *terreiro* was eventually listed by IPHAN in 1984, making it the first Afro-Brazilian space to be added and setting a precedent. IPHAN has since listed other *terreiros.*

A person closely involved with the terreiro told me of a close relative who was a friend of Oscar Niemeyer (1907–2012), Brazil's most celebrated modernist architect. Over a family meal, Nieyemer offered to design something for the *terreiro* free of charge. He knew he had reached such a position in architecture that if he designed something, anything, the site would be more respected for its architectural value. True to form, Niemeyer designed a car park for the Casa Branca together with a square, in honor of Oxum, the "Praça Fonte de Oxum." The squiggly lines are not exactly conducive to easy parking—and the community seems barely to notice the architectural significance—but the project fulfilled its goals.[60]

Figure 85. Casa Branca with Oscar Niemeyer-designed car park. (Photograph by Leonardo Finotti)

Another less common type of *terreiro* is the *Ilé Àṣẹ Òpó Afonjá,* which is like a small city in itself with housing, shops and a school; this type we might call a "nuclear village." *Òpó Afonjá* was founded in 1910 by a group who separated from the Casa Branca. It became one of the bases for the Africanization of Candomblé under the leadership of Mãe Stella de Oxóssi (1925–2018). While *Òpó Afonjá* is centralized on one site, other *terreiros* are dispersed in city, such as the Terreiro de Mãe Menininha do Gantois. Sited in a dense urban area, adjacent to a campus of the Federal University of Bahia, the elements of the *terreiro* are spread around the adjacent urban fabric. It is a "dispersed" or scattered form of settlement. While these *terreiros* represent a range of conventional layouts—islands, nuclear villages, and dispersed urban fabric—they have one thing in common, a green grove. They represent one typology of terreiro, a grove.

Not all terreiros are green, and as the city expands, *terreiros* are inevitably losing greenery to urban encroachment. According to the Centro de Estudos Afro-Orientais (CEAO) mapping of terreiros, 65 percent of terreiros in Salvador have no sacred grove; they are called *terreiros de laje,* or concrete slab terreiros.[61] Increasingly, the only way they can expand is vertically. Sometimes, the space of the terreiro gets sold off to provide income for the community and to ensure its long-term survival. The Terreiro Tingongo Muende, for example, was a green island and sold much of its site for a supermarket. The custodians concentrated an entirely new set of buildings on less than half the previous site. While the architecture is well-designed, the restructuring process meant most of the leaves that covered the site were lost.

The Casa de Oxóssi is in many ways spatially dysfunctional, at least from the perspective of my professional training, biases over efficiencies, and personal space. The Casa de Oxóssi has long outgrown its available space. For the largest festivals and ceremonies, the main religious space of the house is so full that some members are not even allowed inside. There are restrictions on altering the interior religious space, so it cannot be enlarged. Treating the terreiro as a landscape, however, allows us to liberate the rituals from the four walls of the religious space and engage with those outside spaces framed with coconut palms, bamboo, and jackfruit. For example, the

Figure 86. Terreiro Tingongo Muende. (Photograph by Leonardo Finotti)

Águas de Oxalá (Waters of Oxalá) festival in January involves spending the night in the *terreiro.* Members sleep inside in the main religious space (as best they can, considering the hardness of the floor), but the rituals mostly take place outside. Before dawn, around 4 am the community rises—in absolute silence—and, with earthen jars mounted on their heads, process in a line arranged by hierarchy to the well at the bottom of the hill. Each member carries the filled water jar up the steep steps from Oxum's shrine to the main religious space, where it is used to ritually wash and feed Oxalá. (Since I saw the washing of the energy of the air, it has made me think of rain differently.) This festival is incredibly beautiful, but crowded, especially once the older community starts to ascend the hill carrying filled water jars on their heads while younger members are still descending. And once the orishas arrive in the form of trance, they take up even more space with the requisite dancing. In my view, the ritual would work better with some relatively minor spatial adjustments to the outside landscape: an expanded landing here and there and a larger gathering space around the well. A key design question is how the terreiro responds to its audiences, and there are primarily three constituencies to consider: the community of the House, visitors and tourists, and most importantly, the orishas themselves. As terreiros respond to the needs of their contemporary audiences, it would do well do so from a landscape perspective.

Urban Ecologies

As *terreiros* are pressured for space, some of their functions are shifting to public parks which become areas for the performance of rituals. These form a *terreiros-rede* (network of terreiros).[62] There are many public spaces in Salvador being used for orisha devotion, uses that in some cases precede their formal designation as public spaces. Elements of the parks, such as waterfalls and streams, become the focus for the cultivation of energies of nature. Afro-Brazilian rituals associated with energies of landscape features, such as rivers and waterfalls that are harder to find within *terreiros,* are often performed in public. For example, the 75-hectare (or 185-acre) Parque São Bartolomeu, sited in the north of Salvador, was first inaugurated as a public park in 1978 and renovated between 2010 and 2014. Named after Saint Bartholomew, apostle and martyr, the park is also popularly known as the Parque de Oxumaré,

Figure 87. Parque São Bartholomeu. (Photograph by Leonardo Finotti)

Oxum e Nanã. The park is marked by three main waterfalls: one for Oxum, the orisha—or energy—of the river and fresh water; another for Oxumarê, the energy of the rainbow; and the third for Nanã, the energy of mud. A dam dominates the park. Despite only recently being identified as a park, it is one of the last remaining fragments of Atlantic forest that covered Salvador and as such has great environmental importance. It was also the site for several rebellions related to slavery and is thus an important site in the ongoing struggle for racial equality. Not only is the park a public park, and place of national memory, but its natural features, including the three waterfalls, make it a sacred space for followers of Candomblé. A quarter of the park's renovation budget was used for community-related projects, and two new police stations were built to improve security in conjunction with the renovation.[63]

In addition to the Parque São Bartolomeu, other public spaces used for orisha devotion include the Dique do Tororó, the Xangô Stone, and the Casa de Yemanjá. The Casa de Yemanjá is sited near the city on the famed Rio Vermelho route that runs alongside the Atlantic coast. Yemanjá is the orisha of the sea. Linked with a

Figure 88. Casa de Yemanjá, Salvador da Bahia. (Photograph by Leonardo Finotti)

covered walkway to a Catholic church, the Casa de Yemanjá is the site of the famous Yemanjá festival held annually on February 2. Cartesian drawing does not do justice to Afro-Brazilian spaces. Instead, we must find alternative ways to draw them since they are not necessarily understood or experienced through straight lines.

Washing of the Bonfim Steps

A very particular landscape emerges in the city for one day of the year, on the second Thursday after the Feast of the Epiphany on January 6. This urban spectacle is the largest annual public gathering in Salvador da Bahia attracting at least a million people. Almost everyone wears white clothes in honor of the orisha Oxalá and walk all or part of the eight kilometers (five miles) from Salvador to Bonfim. For some it's a day of religious devotion, for others an opportunity to drink beer from the early morning. The day, which is a public holiday in Bahia state, begins with an early morning Mass at the church of Nossa Senhora da Conceição da Praia (Our Lady

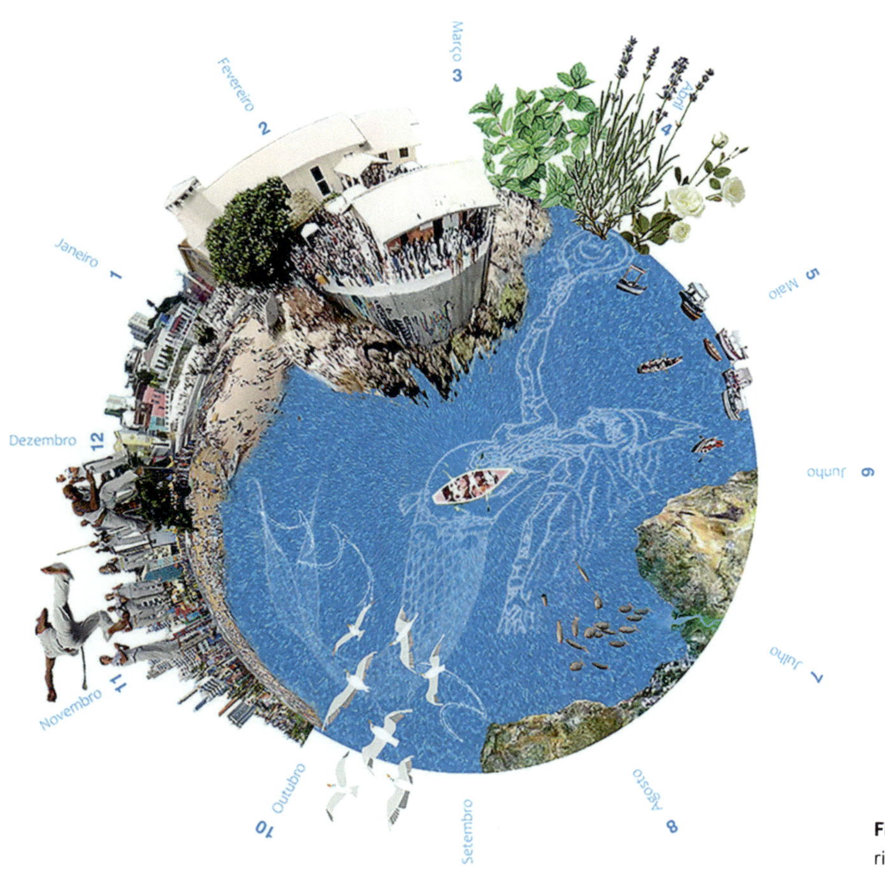

Figure 89. The annual cycle of rituals in the Casa de Yemanjá. (Drawing by Michele Turrini)

of Conception of the Beach) beside the market in Salvador where enslaved peoples were formerly traded. After the Mass, the priest and various iyaorishas, babaorishas, and other religious leaders together recite the Our Father. The procession, carrying a statue of Our Lady, follows a direct route to the church of Nosso Senhor do Bonfim (Our Lord of the Good End). The hill of Bonfim is noted for its special air quality, after all, it is dedicated to Oxalá, the orisha of the air. A couple of days after the Bonfim festival in 2019, I met a priest who proudly reported that more than 2 million had participated in the event.[64] The steps of the church are washed with aromatic water and the same water was used to bless the faithful. While the bishop leads prayers from the balcony above the church, the church remains closed that day. A recent square designed by Sotero Architects in front of the church incorporates the symbol of Oxalufã in the paving pattern. The slow and thoughtful Oxalufã is one of the two paths of Oxalá in Brazil, the other is Osogiyan, the warrior.

Figure 90. One million people wearing white. (Photograph by Uiler Costa)

Learning from the Terreiros

So what? How is a process of gathering leaves useful for design? This may lead us to ask how a landscape architect might design for an Afro-Brazilian religious community. Are there other forms of landscape architecture—the revealing of knowledge for knowledge's sake—for instance? Or the critical reflection on whether an intervention is justified or not? The understanding of the terreiros as part of a wider urban ecology? My fieldwork in the terreiro has led to some very minor design interventions, and prevented others, but the main contribution (at the time of writing) is this chapter and situating Afro-Brazilian spaces within a landscape-architectural narrative. The study led to a conference, Sacred Groves and Secret Parks: Orisha Landscapes in Brazil and West Africa, as well as a study on the urban ecology of *terreiros.* These various initiatives were made possible through my embodied engagement in the field.

While there has been growing scholarly interest in the African diaspora, the so-called Black Atlantic,[65] and the growing field of Afro-Latin American studies,[66] there has been relatively little attention to the flows and crossed perspectives on spatiality, landscape architecture, and urbanism. As we move between sacredness and secrecy in the inhabitation and understanding of Afro-Brazilian sacred groves it is important to consider how landscape architecture can learn from these spaces. Some designers have worked with the knowledge from Candomblé to design new landscapes. For example, the Rio de Janeiro-based landscape architects Embyá, headed by Duarte Vaz, designed an ethnobotanical garden in Salvador. Fieldwork is essential, Vaz says. "Landscape fieldwork is what feeds the entire (design) process. Without it, the work becomes a generic project without a soul."[67] The process included participatory engagement among Candomblé leaders who worked with Embyá to choose the places for the different orishas' gardens. Elena Geppetti, a landscape architect with Embyá, is a Afro-Brazilian ethnobotanical scholar. The architect Vilma Patricia Santana Silva developed a deeply immersive fieldwork-based design methodology in which she, as architect, consulted the orishas through throwing cowries, a process of divination.[68] So immersed did Vilma become in her fieldwork that she was initiated into Candomblé. The EtniCidades research cluster at the Federal University of Bahia (UFBA) likewise adopts a deep fieldwork ethos: field-based knowledge is essential without a literature to work with.

Afro-Brazilian *terreiros* are a token of an emerging type of environmental preservation, oscillating between private and public urban spheres, between the sacred and the secret, and their study can help in formulating a theoretical argument about the role of third-sector religious organizations and other nonhuman agencies in the future of parks. (Third-sector organizations sit between the binaries of public and private, strictly neither one nor the other and include religious organizations, charities, NGOs, and other nonprofit-making, nonstate entities.) Numerous third-sector urban spaces, especially those that belong to religious organizations, could have a public use. The *terreiros* reflect the coming out of Afro-Brazilian religions to the public domain and help in locating landscape architecture as a multidimensional form, expanding upon, and challenging the Western canons of the field.

Reflections

This chapter described the insights I gained from doing slow ethnography—how I was able to structure my research in such a way that I could return to a particular site, an Afro-Brazilian *terreiro* in the city of Salvador da Bahia, for short periods of time over some years. Other points in the chapter—such as what the *terreiros* themselves can offer landscape architecture—are insights gained through this specific method of landscape fieldwork, or "gathering leaves." One of the threads running through the chapter is, when do you intervene as a designer, and when do you sit back and stay silent, listen, learn, and work? We can see how embodied engagement—in mundane everyday activities like cleaning and more ecstatic, exceptional moments like trance—all help to understand the value of *terreiros* and to see them as offering a radical re-envisioning of what it means as a designer to engage with public spaces and their communities.

Terreiros are fundamentally based on a respect for nature. We see how much care is attached to leaves, and their centrality for trance and transcendence. The *terreiros* are inherently productive spaces, cultivating plants and animals, not to mention the energies themselves although, admittedly, many of the productive functions are outsourced today. The *terreiros* have deep social structures and entanglements that extend far beyond the state of Bahia.

Landscape architecture can learn from *terreiros* at the scale of the city—as secret parks and providers of ecosystem services—but perhaps more so at the scale of the *terreiros* themselves. *Terreiros* are inherently political spaces. On a basic level, some energies like each other, while others repel. The spatial logic of *terreiros* is something deeply political. Social hierarchies and personal histories also help to give form to the *terreiros.* It is within the internal politics and dynamics of the *terreiros* that we can find clues for what they can offer to the city and to concepts of public landscapes.

In Brazil, where architecture and urbanism dominate over landscape architecture, *terreiros* are often considered architecture. *Terreiros* are landscape architecture, just not designed by landscape architects. In this sense, they escape our traditional professional boundaries and discourses. *Terreiros* provide housing for the less privileged, and a home for people with no home. They are spaces for the disadvantaged, descendants of enslaved peoples, members of the LGBTQ+ community, the poor, and oppressed. Many members are considered on the margins of society. Social structures in the *terreiros* are inverted, those on the periphery are empowered, and privileged professors (like me) are disempowered. Too often, landscape architecture is for the privileged. *Terreiros* touch those who need empowerment.

To enter a *terreiro,* one leaves behind the world outside. Imagine a landscape which is at once of the city and is also another world requiring a formal act of transition. You cannot simply enter from the street but must first be cleansed. And then work. The act of working in the *terreiros,* and doing whatever one can within one's own limitations, makes everyone an active participant. What if instead of being consumers of landscape, we become active participants in the management of a landscape? It is not possible to consume the facilities of the *terreiro* without giving something back: one must work, and "gather leaves" in a prolonged process of embodied engagement. Gathering leaves makes everyone a participant observer: every member of the community learns through landscape fieldwork. To begin to engage with landscape architecture in this way requires an acceptance of the other, rather than shaping of the other to ourselves. If we were to do that as a field, what a wonderful world we would shape.

6

THICK PRESCRIPTION

A Framework for Landscape Fieldwork

Landscape architects are often confused with gardeners. And it's true that landscape architects can and do design gardens, such as Shute in the English countryside. But more often, landscape architects work with the elements of a garden over a larger area—across scales, temporalities, and landscape types. Landscapes are complex aesthetic, environmental, and social constructs, and the landscape fieldworker must employ a range of methods and work with multiple elements to understand them. They may work with living plants, shrubs, and trees, or inert materials such as soil, wood, stone, and concrete. Landscape architects pay close attention to color, form, scent, texture, and volume, as well as biological, geological, hydrological, and other environmental processes. Importantly, landscape architects also work with the element of time: landscapes are dynamic and often come into maturity long after the lifespan of an individual landscape architect or owner. And the essential element in landscape architecture is people. If we don't design for people, who are we designing for? And if we don't understand the people we are designing for, how can we design for them?

The Landscape Architecture Accreditation Board in the United States outlines two primary principles for the profession, the first of which is people-based: "to protect the interests, well-being, and safety of people and communities, including future generations; and to safeguard the health and resilience of natural systems, ecosystems, and non-human inhabitants."[1] It is telling that they put people first. As landscape architecture grapples with ever more complex environmental, socio-spatial, and urban issues, the field must also become better equipped to contend with the human dimensions, including behaviors, perceptions, and values. Yet landscape architecture education and practice tend not to include the social dimensions, focusing instead on a narrower environmental and aesthetic framing of the field. As I have demonstrated throughout this book, and elsewhere,[2] it is ill-advised to arrive at a technical solution through purely technical means that neglects the interrelationships, and blurred boundaries, between nature and culture, and humans and nonhumans, that many scholars have written about.[3]

Such questions become especially salient when we work with complex and densely inhabited urban sites, what Anne Whiston Spirn calls "the granite garden."[4] Shute, as we've seen, is a garden set within a mostly rural landscape; it's a site surrounded by a network of small villages, located a short drive from the city of Shaftesbury and the London-Exeter line train station at Tisbury; about two hours' drive from London. But most landscape architects do not get commissioned to design elaborate gardens like Shute. Instead, we are increasingly called upon to design public spaces, engage in urban redevelopment and climate change adaptation, which requires comprehending the myriad assemblages of relationships in and between landscapes, people, politics, and practices—the urbanism of landscape.

In recent years, landscape architects have rightly been reclaiming the city as a site of expertise. While one history of the profession claims its origins in the city—the canonical example of Olmsted and Central Park comes to mind—over the years the profession has somehow subcontracted or delegated that urban responsibility to urban designers and city planners, sometimes even artists. We did this by letting go of the human aspect. When we disregard the human dimensions of landscape architecture, we cannot fully realize landscape architecture's political potentials. When we claim the human dimensions of landscape architecture, we can realize landscape architecture's political missions.

Ian McHarg's *Design with Nature* transformed landscape architecture and rightly foregrounded the field's ecological potential but neglected people.[5] For this reason, *Design with Nature* has done a disservice to the field. James Corner, Laurie Olin, Anne Whiston Spirn, Frederick Steiner, and many others, have built upon McHarg's legacy and advocated for the human dimensions—that he left out—through their texts and projects. One thinks of Corner and MacLean's *Taking Measures across the American Landscape,* Olin's *Breath on the Mirror,* Spirn's *The Granite Garden,* and Steiner's *Human Ecology.* There have been shifts such as landscape urbanism, and ecological urbanism which claim the city as a site of focus—and by extension the human—and place the architecture of landscape at the core of urbanism—where it should be.

Meanwhile, Félix Guattari's short book, *The Three Ecologies,* offers a robust framework for understanding landscape's urban and, by extension, human agency. Guattari, who trained as a psychiatrist, has a very different academic genealogy to most landscape architects, but his words speak directly to us. Guattari reasons that we can only bring about more sustainable cities if we combine environmental, social, and subjective ecologies. Often these "ecologies" are considered separately with the environmentalist, the social scientist, and the, say, artist doing their own thing. I have found, unsurprisingly perhaps, that when we combine these three ecologies, and take the wider view that Guattari advocates for, it often results in what might traditionally be termed landscape architecture. Landscape architecture has the means and know-how to bring about the multidimensional ecological future that Guattari calls for. As a result, the three ecologies provide a helpful metric with which to measure landscape architecture. Often design projects score very high in one or two of these ecologies, but rarely in all three.[6] They remain a challenging goal to aim toward.

Roberto Burle Marx, who would not qualify as a landscape architect under today's stringent professional requirements, presciently saw this shift toward grappling with complex urban problems as the way forward for the field. Burle Marx envisioned a future where the neon signs and traffic lights of the city could be the flowers of the future gardens. In a lecture entitled "The Garden as a Form of Art," he asserted, "In New York, or even in Rio de Janeiro, the neon signs, the advertising posters, the traffic lights, the lighting of parkways—these are not problems that can be ignored. From these is born a new aesthetics."[7] Burle Marx challenged landscape architects to take on these new elements. The "garden has left the hand of the gardener," Burle Marx declared, "who although he may have had great knowledge of plant material was not solving the problems of the gardens of the modern city."[8] In this one statement, Burle Marx recognizes the unique skills of the landscape architect, dealing with plant material and consequently materiality, texture, volume, color, and light, and applying these to the hardness and complexity of the urban condition. Burle Marx maintained that it is essential for the designer, whom he terms "the organizer of urban space,"[9] to grasp human behavior, including "social structures and group tendencies."[10] At the same time, Burle Marx recognized the limitations of the skills traditionally associated with landscape architecture for engagement with the complexities and the human dimensions of cities. Consequently, Burle Marx saw a need for sociologists and economists among design teams.[11]

As the field has grown, more practitioners and scholars have become convinced not just of landscape architecture's limitations but also the need for a more multifaceted approach to understanding landscape: a reimagining of the field. Like Burle Marx, the Danish landscape architect Stig L. Andersson, for instance, identifies the need for integrating a wider array of disciplines in the architecture of landscape, a practice he calls "the new nature." The new nature recognizes its own artificiality in the face of the destruction of natural environments and construction of new ones. Andersson's Copenhagen-based firm, SLA, engages planners, architects, botanists and, notably, has several full-time anthropologists in the office. Landscape architects are joined by these other fields in the construction of the "new nature."[12] SLA recognizes the essentiality of engaging with the human condition when engaging with the city. Other firms, such as Kounkuey Design Initiative (KDI),[13] also invest deeply in human ecologies. Of the thirty or so people working in KDI's Nairobi office, one-third are involved in public engagement both in terms of learning from the various publics and explaining a project to them. KDI recognize that there is a complex to-and-fro of information, and this is one of the reasons their work is so impactful.[14] Meanwhile, PARKKIM, a Seoul- and Boston-based landscape and architecture firm founded by Yoonjin Park and Jungyoon Kim, engages in what it terms "alternative nature." With this term, PARKKIM refers to an artefact, or artefacts, that offer the experience of (so-called) nature in urban settings, and consider the lost functions of nature as an aid in climate change adaptation.[15]

By focusing on urban sites, landscape architects can claim their promise to effect wider social and spatial change.[16] In the previous chapters, I've advocated for landscape fieldwork as offering invaluable ways for landscape architects to engage with and learn about the city and other complex landscapes. Let us now turn our attention to what it is about landscape fieldwork that is distinctive in some way. For if the city of the future is to be an amalgam of three ecologies—the environmental, the social, and the mental in a broader Guattarian sense—then we need the skills to identify, understand, and work with these three ecologies in an expanded field of landscape architecture.[17] Landscape fieldwork, as a core practice of landscape architects, has multiple intellectual genealogies and draws from diverse academic and professional fields. The exact landscape fieldwork methods and their deployment for a given project must be determined on a case-by-case basis by the site and its limits, the climate, and the fieldworker's experience and preferences, among many other factors. The fluidity of approaches can make landscape fieldwork seem mercurial, but it doesn't need to be so. There are four fundamentals to a landscape fieldwork approach.

Four Fundamentals for Landscape Fieldwork

In the previous chapters and cases, I have introduced different approaches to landscape fieldwork across diverse landscapes and temporalities. We've seen that landscape fieldwork has no singular application; rather, the practice is site-specific and depends on the landscape—including its human and nonhuman inhabitants—the project, and the fieldworker. In the introduction, and based on the range of fieldwork across fields, I outlined four fundamentals for landscape fieldwork. Phrased as four, nonlinear instructions, these are

1. Mix measures
2. Immerse your body

3. Contrast media
4. Critically imagine

Let's look more closely at each of these interrelated steps and why they are so fundamental for landscape fieldwork.

Mix Measures

Landscape fieldwork measures across space, time, and context. Landscapes are physical; they are composed in space. Landscapes are social; they are constructed through an assemblage of relationships. Landscapes are relational: the mutual shaping of people and place.[18] In many ways, landscapes are the physical manifestation of social relations; the landscape fieldworker unearths relationships, and the landscape architect gives them physical form. Landscapes are temporal; they grow and change over time. Landscapes are sensual; not just appealing to the visual but sounds, smells, and touch. The work of the landscape fieldworker must therefore be measurable in space, time, and human dynamics, including feelings and senses. The sensorial, social, and temporal aspects of fieldwork are well addressed in other fields, such as anthropology, art, and ecology; the spatial aspect is what makes landscape fieldwork so distinct. Space is composed and constructed through smells, sounds, and politics and not just its physical boundaries. Landscape is both quantitative and qualitative and therefore requires both forms of measurement.

The cultural geographer J. B. Jackson summarizes the definitions given in most dictionaries, which tell us a landscape is a "portion of land which the eye can comprehend in a single glance."[19] He then cautions, "Actually when it was first introduced (or reintroduced) into English it did not mean the view itself, it meant a picture of it, an artist's interpretation." The landscape ecologist Richard T. T. Forman echoes Jackson when he defines the extent of a landscape as "the object spread out beneath an airplane window,"[20] and goes on to describe landscape as an area of countryside or city larger than an ecosystem and smaller than a region.[21] That might well be the case from a definitional perspective, but the practice of landscape architecture demands a dexterity across scales. While the impacts of our work at various scales may differ, as Carl Steinitz rightly points out, it's still not a reason not to work across scales.[22] In fact, one needs to be aware of the implications as we work towards designing at the global scale.

Landscape fieldwork is not tied to any one scale but understands scale in context. The landscape fieldworker works across scales. The cases in this book ranged from a modest village square to an archipelago of 365 islands that extends over 190 kilometers (120 miles). They both employed a form of engagement that extended from the residents to local and national government. Even though the square was something I was designing at home, the process in Ireland brought me to the centers of power in Dublin and London (chapter 1). While working with the center of power in Nassau, The Bahamas (chapter 4), we worked with the dispersed island-based communities. The 365 islands of the Exuma archipelago can only be understood in relationship to other islands, and to the archipelagic nation of The Bahamas. Like the small chicken coop in Exuma, landscapes are relational, and what affects one part can have a larger spatial impact.[23] Landscape fieldworkers work from very small sites to the very large, from balconies and gardens to entire regions to the scale of the world, and beyond. Climate change is a global issue and landscape architects can do so much to address it—doing so requires us to work from the domestic across the global scale, and more. Because of its interconnectedness and relationality, landscape architecture is inherently prototypical. In chapter 2, I introduced the idea of a prototype having a process as well as a form, and of a prototype as a hybridizer and proliferator. Landscape architects work with processes in bringing form to reality, and these forms can be generative of new forms and programs.

Landscapes are tangible and speculative, earthy, earthly, geographical, transcendent, and topographical entities. Landscape fieldworkers are interested in recognizing and crafting how *places* become *spaces* through design, meaning, and activity. Michel de Certeau in *The Practice of Everyday Life* tells us that place is to space "like the word when it is spoken."[24] As we saw in the prologue to this book, Certeau defines space as "a practiced place."[25] Thus, places become spaces because of practices of use, inhabitation, and memory—and their potentials. These are the mixed measures that the landscapes fieldworker addresses.

The landscape fieldworker will use measurement tools such as a tape, theodolite, level and rod, 3D-scanner, GPS, and even the ageless art of pacing to understand a landscape's physical measure. The fieldworker might gauge quantitative aspects such as length, amount, weight, but also a landscape's temporality—time-lapse photography, satellite imagery, and oral histories are all important tools in this regard. The landscape fieldworker might take stock of geological time or botanical time by examining stones or rocks, or indeed the trees, shrubs, and plants that comprise a landscape. The landscape fieldworker will quantify—and qualify—the cultural and projective dimensions of a landscape too.[26] In *Taking Measures across the American Landscape,* James Corner explores landscapes' multiple measurements by juxtaposing his maps with Alex MacLean's aerial photographs to interpret human practice on the land in much the same way as an archaeologist, ecologist, geographer, or planner reads the landscape from aerial imagery.[27] Likewise, Anuradha Mathur and Dilip da Cunha's inspiring maps of the Mississippi Delta include not just the physical landscape but the nonphysical landscape too, such as the songs that were sung by enslaved peoples as they worked on the cotton plantations.[28] While the landscape fieldworker will measure time, perhaps geological time, or botanical time in terms of the trees, shrubs, and plants that comprise a landscape, at the core of a landscape's measure are the temporality of day and night, the seasons, the human lifespan, through which human's measure and make sense of the world.

From an environmental perspective, a landscape needs to have heterogenous qualities to foster biodiversity and a healthy ecosystem.[29] From a prescriptive capacity the landscape fieldworker anticipates future measures. In the process of thick description, the landscape fieldworker will focus on a landscape's place within a larger ecosystem and social and political region as much as the space within the confines of a garden, such as Shute, an amalgam of the classical and romantic into the "cosmic." When it comes to measurement, Guattari's three ecologies—the environmental, social, and subjective—are helpful because they don't separate the world into binaries of hard and soft, or qualitative and quantitative. Guattari's three ecologies offer a useful multipronged instrument for measuring a landscape.

Immerse Your Body

Embodied engagement is, in some ways, like an actor getting into character. As actors begin to engage the mindset of the individual they portray, there is an element of becoming the other. Actors might learn new mannerisms, skills, and accents in this very act of trying to experience the world through others' eyes. Landscape fieldworkers, similarly, will immerse themselves in the realities of their interlocutors to understand their perspectives and ways of life. Usually, the goal is to understand various phenomena through their study *in situ.* Fieldwork unearths patterns and relationships that might have gone unnoticed otherwise. In the process, the fieldworker themselves need to be open to transformation.

Like the ethnographer, the landscape fieldworker aspires to what the anthropologist Subhadra Mitra Channa refers to as "embodied engagement."[30] Channa describes embodiment as "the entire process of formation of the self and identity from within the bodily location."[31] In terms of fieldwork, one's embodied presence in the field influences the way others engage with you, as well as one's own gaze, which is "turned outside but within one's body."[32] Channa attributes the concept to both the Western philosophical discourse of phenomenology following the philosophy of Edmund Husserl and others, and from Hindu philosophy, where mind and body are seen as unitary.[33] The unified mind and body is the landscape fieldworker's central tool.

When I lived in Ireland, I knew how to be Irish. But when I lived in Bahrain, I had to learn how to be a Bahraini. I took Arabic classes, and courses in Islam. I observed local dialect, codes, and behavior patterns. In much the same way as languages are best learned through immersion, I learned how to be a Bahraini through immersing myself in the everyday life practices of my interlocutors. I learned about the meaning of *shlawnak* by having dinner with my friend Isa. I ate meals with my interlocutors and, when I ate alone, I ate at the same few restaurants. I got to know the staff and customers as well as the menus and food sources. The landscape fieldworker has strategies for becoming, and for setting up the possibilities for chance encounters to happen. Walking to talk through random encounters offers an immersive form of random sampling, not unlike throwing beans on a map. Walking exposed me to the multidimensionalities of landscape and its myriad smells, textures, and colors. Walking exposed me to people, and especially people I would not otherwise have met.

The miniscenario method is important because it results in a form of miniethnography. Landscape fieldworkers immerse themselves in the landscape by getting to know its two skins. They are interested in the hidden processes of a landscape—social, emotional, spiritual—as much as in its visible attributes. Immersion in a miniscenario—a small glimpse into the processes flowing through a place in time—can be as intense as immersion in a sacred bath of herbal leaves in an Afro-Brazilian grove. The immersive experiences described in the preceding chapters all involved different temporalities. The first chapter in this book is based on a lifetime's immersion at home and the depth of insights and relationalities that this brings. The second chapter was based on the fleeting immersive experience of the miniscenario, which reveal operational fields and take their strength from aggregation. Chapter 3 described four stages of my Bahrain fieldwork in which I learned how to be Bahraini (remote and preliminary fieldwork, acclimatization, becoming; and writing up and fact checking). These stages were speeded up and dispersed among a collective (chapter 4) in The Bahamas. The four processes were slowed down in Bahia, Brazil (chapter 5), where through a process of embodied engagement, I unexpectedly gained the experiential knowledge that propelled me beyond the narrow limits of my inquiry. At the time, I felt the knowledge gained from that immersive experience transcended five years of reading books.

As the landscape architect Jala Makhzoumi pointed out to me in conversation, when landscape fieldworkers immerse our bodies, we also immerse our entire spirit and soul. Like the transcendentalists, the landscape fieldworker recognizes landscapes' spiritual dimensions. When traveling to and from field sites, I often visit airport chapels and prayer rooms. They are fascinating to me because they are intended for people of multiple faiths, and none, to engage with the transcendent. This engagement can be fleeting, as one is rushing to catch a flight, but they can be nonetheless intense. Because of the multiple audiences, these spaces are often designed to be devoid of religious symbols so as not to privilege one faith tradition over another or offend anyone. In lieu of crosses, crescents and Stars of David, such spaces will often hang a photograph or painting of a landscape. The implication being that landscapes allow contact with our deeper selves in ways that transcend religious dogma. The landscape fieldworker is open to the transcendent as an integral aspect of a landscape.

Increasingly, the digital age we inhabit brings a set of challenges and opportunities for the immersive aspects of landscape fieldwork. For sure, it's possible to immerse oneself in the Internet, the merits and limitations of remote fieldwork to complement the embodied engagement of field-based fieldwork are challenging preconceived notions of the value of immersion. Artificial intelligence, alternative realities and second lives, Google Street View, and satellite images, bring many opportunities to expand upon the tools of the landscape fieldworker. Yet, as Stig L. Andersson, so astutely observes, "the more digital things become, the more you need to feel."[34] Anne Whiston Spirn would agree: she writes, "Landscape, as literature of lived life, must be experienced in place to be fully felt and known."[35] This is not to dismiss the digital, though. The anthropologist Daniel Miller, in a YouTube video created during the COVID-19 pandemic, argued that if we make a distinction between virtual and actual fieldwork, we are not thinking anthropologically.[36] This is an excellent point, for I would conceptualize doing remote fieldwork as a complement to traditional immersive strategies, rather than a replacement. The digital does not replace the analog, the virtual does not replace the actual; they complement and inform each other. Rather than being replacements, they are important aids to the immersive, embodied experience, of being in the field.

The depth of the immersion is often more important than the duration of the immersion. One intense month in the field can be more valuable than a year spent poorly. And I have found that the more intense the landscape fieldwork, the more careful the design authorship. This may suggest that landscape fieldwork heightens our critical awareness. It helps us to understand whether a project is necessary or can be built in another way. Landscape fieldwork allows us to engage our body and immerse our spirit to critically understand a landscape and its future possibilities.

Compare Media

The landscape fieldworker uses multiple modes to represent the world, its first and second skins, observed and imagined. As well as writing fieldnotes and sketches, landscape fieldworkers will use their tools of sectioning, transecting, and perspective to understand a landscape through drawing. Sections through a hillside, for example, can help the landscape fieldworker to understand topography, soils, hydrology, and geology but also map movements of people in space and time. As

Jellicoe observed when working on the drawings of the Italian Renaissance gardens: "The effect was staggering. The drawing that slowly emerged revealed that the substance of the beauty of a garden could be transmitted to paper."[37] Jellicoe's later drawing style, which he developed in his eighties—using dots to illustrate depth and shadow—was a form of meditative practice. He would spend Saturday mornings in his home office with the pen on paper. His wife, Lady Susan, would ask, "Are you dotting again?"[38] The human element is essential, even though it can be difficult to draw and map. As with written interpretations, maps don't just reproduce the world; they construct the world.[39] We draw not just to represent, but to find.

There is a strong link between forms of representation, and the eventual appearance of a project. Designers are often limited by the limits of the medium. Had a project been represented in another way, the eventual form would most probably be different. Albena Yaneva makes this point in her insightful ethnography of the Office for Metropolitan Architecture (OMA) in Rotterdam. Through the immersive act of fieldwork, Yaneva shows how OMA's chunky buildings bear an uncanny resemblance to their blue foam study models.[40] Meanwhile, the act of notation helps one to understand and construct a space. Tim Ingold tells us: "The basic principle is that notation is anything that can help transfer movement from one medium into another: for example, from something written on paper to a musical or a dance performance."[41] If we draw in a certain way, say with a vector-based computer program, then the resulting project will likely bear a physical resemblance to that drawing. Zaha Hadid's radical forms were readily attributed to the computer programs that allowed them to be drawn. The architectural historian Robin Evans, relatedly, writes about the role of drawing in architectural design;[42] what we draw influences what is built.

This is important because the ways we understand landscapes and represent them in our designs ultimately influences built forms. Landscape architects attest to this linkage again and again: that what they understand about a site, influences what is built. As Laurie Olin and many other landscape architects have attested to me in conversation, there is a strong link between forms of landscape fieldwork and the physical form of an eventual project. Landscape fieldwork unravels the kind of experiential knowledge that designers need to be better designers. The landscape fieldworker works with a variety of environmental and social processes, which requires an intimacy with a site, the objects of observation, and their interactions. The intimacy may come from quantitative data as much as it comes from living among a community and learning its language, codes, and patterns of behavior. By observing, recording, and interpreting, the landscape fieldworker is creating the generative potentials of a drawing or a text.

Part of what makes this immersion into a practice is the keeping of detailed notes—through writing, painting, photography, audio recordings, sketches, video—and then interpreting those notes. The rule of thumb I've mentioned—for every hour spent in the field, the fieldworker needs four hours to interpret—ensures that adequate time is spent in reflecting and analyzing what is experienced. Landscape fieldwork without interpretation is like having words without a sentence, a string of unlinked signs and symbols. Landscape fieldworkers decipher what they see and give it form. They work, listen, observe, take in what is happening around them, and through quiet reflection eventually add the leaves together to make up a plant. Admittedly, this 1:4 ratio is difficult to follow, given the financial and temporal limitations of a design process, not to mention the pressures of time constraints in the field. The temptation might be to skimp on the interpretation, but the interpretation is the most important part. And it is never neutral. Ingold tells us that the act of interpretation is shaped by the worldview of the perceiver: "Any act of description entails a movement of interpretation. What is given to experience, in this mode, comprises not individual data but the world itself. It is a world that is not so much mapped out as taken in, from a particular vantage point, much as the painter takes in the landscape that surrounds him from the position at which he has planted his easel."[43] Landscape fieldwork is like painting a picture: only the artist knows when the painting is complete. The centrality of the fieldworker's own observations and feelings is fundamental to the practice of landscape fieldwork. Jellicoe also described his work at Shute as being akin to a portrait painter. He said: "When I design a garden for a particular client, it is rather like painting a portrait; you have to get inside the clients' minds and translate their wants through your art."[44] In this sense, the designer is not giving the client exactly what they ask for listening to them and trying to understand *why* they're ask for something. The job of landscape architects is to "translate" and interpret what they hear. The critical aspect here is the critical reflection, and imagination, of the landscape fieldworker.

The assumptions one unravels from landscape fieldwork, ultimately are figured in modes of representations. There are specific limits of representation that we must acknowledge if we follow this paradigm. The media involved bring with them certain assumptions. For example, in the Diamond project in Ireland, the conventions of plans and sections were of limited value as I found the public did not understand them so well. Instead, the public responded to the three-dimensional model and drawings and photorealistic images showing the site before and after the proposed intervention. In The Bahamas, we were told that the public would not read our one thousand pages of text. Instead, they wanted to have oral modes of communication to reach the people, which we produced as part of a larger exhibition at the National Art Gallery of The Bahamas. In drawing, we were influenced by Anuradha Mathur and Dilip da Cunha, who, for example, represent water across terrain in novel and interesting ways, rather than following the graphic convention of representing water with a line, which is never the case on the ground. When you shift the medium from one form to another, say from writing to drawing, and then compare them it pushes you toward more design possibilities. The literary critic Y. M. Lotman, in his "Text within a Text," writes about intermedial textuality.[45] A soundscape might be related to a model, a movie, or a poster of a movie. These juxtapositions increase the possibilities for interpretation and—for the generation of new modes of thinking—understanding, and designing. Lotman writes, "In an overall cultural system, texts fulfill at least two basic functions: to convey meanings adequately, and to generate new meanings."[46] The generative possibility of the text brings us to the fourth category: to critically imagine.

Critically Imagine

The landscape fieldworker engages with critical landscapes in critical ways, imagining possible futures for them. Some of these landscapes are overlooked and outside the formal canons of the field. They are critical because they directly address the misaligned hegemonies of landscape architecture, bringing landscape architecture to minorities and, critically, minorities to landscape architecture. The *Stanford Ency-*

clopedia of Philosophy, quoting the critical theorist Max Horkheimer, writes that "a theory is critical to the extent that it seeks human emancipation from slavery, acts as a 'liberating influence,' and works 'to create a world which satisfies the needs and powers' of human beings."[47] The landscape fieldworker works within such an agenda that flattens perceived hierarchies. Horkheimer further outlines three criteria for a critical theory: it must "explain what is wrong with current social reality, identify the actors to change it, and provide both clear norms for criticism and achievable, practical goals for social transformation."[48] In this sense, a critical theory is projective and aimed at changing the future.

Critique is not just central to a reflexive design practice;[49] it is a creative act, and as Jacky Bowring makes clear, it takes place at every stage of a design process.[50] Whether in school or in a professional office, designers rely on peer review as a form of critique, from conceptualization, to design, construction, and the afterlife of a project.[51] This peer review is both about finding faults and about finding merits. In doing peer review, the critic is engaged in "critical thinking," or "careful thinking directed to a goal."[52] The landscape fieldworker critiques through the immersive knowledge obtained from the field, from recorded experiences. drawings, sketches, as well as interviews, formal and informal. Criticism extends beyond the actual fieldwork to the effective sharing of the research.

Anita Berrizbeitia writes that criticism works internally within the field of landscape architecture to contextualize built work historically and broaden disciplinary frameworks and forms of practice; and criticism also works externally in public discourse. Berrizbeitia explains that the potential of criticism is to "augment the legibility of the nature-landscape boundary to expose that which lies behind the making, and the distribution, of landscapes: who is included, who is excluded, who pays, who benefits, who and what are represented and who and what are not."[53] Following such careful goal-directed thinking should mean that landscape fieldworkers do not accept received knowledge without questioning it. Neither do they uncritically accept future predictions.

The critical imagination is at the core of each of the chapters in this book. Chapter 1 described a design project that was subject to the minute scrutiny of village residents, to local government through to the national government and president, not to mention design authorities such as Liam McCormick and Sir Geoffrey Jellicoe. The project was accepted by the community through a complicated process of critique, and part of the project's success was due to the fact it was not positioned outside the formal hegemonies of the state but in parallel with them. It was a bottom-up process, tempered by a top-down strategy. The CHORA Urban Gallery (chapter 2), has many levels of critique built into it, none more central than the scenario games. Speculation on what a project might be, and the form it will take, can be wild and untethered, yet is grounded and tempered through the scenario games, which critically evaluate a project and its possibilities for change before the action plan is drawn up. In my landscape fieldwork in Bahrain (chapter 3), I was challenged to find a way of writing that would respect the numerous social groups without alienating one group over another. I chose Bahrain for its extreme conditions with lessons for more temperate climates. I had Bahrainis read my ethnography before publication: another dimension of critique. The fifty-two fieldworkers dispersed among The Bahamas (chapter 4) were critical agents engaging in landscape on their interlocutors' terms. The work was critiqued by the government and islanders alike. And in chapter 5, in Bahia, in the Afro-Brazilian sacred groves, fieldwork allowed me to question the need for a major intervention. The Afro-Brazilian groves are so far removed from the Western canon that they were illegal till the latter part of the twentieth century.

"Criticism is itself an art," declares Oscar Wilde: "The critic occupies the same relation to the work of art that he criticizes as the artist does to the visible world of form and colour, or the unseen world of passion and of thought."[54] Thus the role of the critic is to imagine. "Criticism is essentially an optimistic enterprise," declares Marc Treib.[55] The philosopher James Grant tells us that "imaginativeness is a propensity to think of unobvious achievements."[56] The landscape fieldworker as artist, critic, and designer, is deeply involved in the act of imagination and constantly aware of the possibilities for change, imagining future scenarios.

The critical imaginative, projective, design dimension is crucial for landscape fieldwork. Like landscape, and imagination, the definition of design can be hard to pin down. The architect and urban designer and former dean of Harvard Graduate School of Design, Peter G. Rowe, writes, "Design appears to be a fundamental means of inquiry by which man realizes and gives shape to ideas of dwelling and settlement."[57] The political scientist Herbert Simon is often quoted when a definition of design is needed. Simon tells us, "Everyone designs who devises courses of action aimed at changing existing situations into preferred ones."[58] Landscape as the mutual shaping of people and place comes closer to recognizing the impact of design on us. The critical imagination is about anticipating alternative futures, and more precisely preferred futures, and their impacts on human and nonhuman audiences, not to mention the field at large. The anthropologist Robert B. Textor makes seven points about what he terms "anticipatory anthropology." Textor 's first point is that when we do anticipate the future, it is often correct: "Our anticipations of the future do, after all, often turn out to be essentially correct," he writes.[59] (Beware of the current warnings over climate change.) His fourth point is that the designer should visualize the future: "The best way to realize a preferred future is first to visualize it—in concrete enough terms to provide the motivational basis for appropriate action."[60] The designer is concerned with imagining these preferred futures in concrete ways that can lead to physical projects. He goes on to warn, "An individual, community, or society that fails to visualize its preferred future risks being seriously dissatisfied with what it gets when the erstwhile future becomes the present."[61]

Speaking about research, practice, and pedagogy, the South African urban scholar and director of the African Center for Cities in Cape Town, Edgar Pieterse, identifies "the equal importance of rigour, grounded engagement, criticality and the transformative joys of inquiry into the unknown, unseen and misunderstood."[62] Landscape fieldwork offers similar possibilities for the critical imagination. The landscape fieldworker approaches a field site with an open mind, always asking how things could be different, and usually with a project in mind from the outset. The landscape fieldworker's engagement with a landscape is descriptive in the sense that it is about a detailed, "thick" understanding of that landscape; and prescriptive in the sense that is open from the outset to build on the description to carry out an act of design. Thick prescription is rooted in the ecologies and operational fields embedded in a landscape, and the landscape fieldworker moves between description and prescription, often inhabiting the space between them:

Toward Thick Prescription

Because landscape fieldworkers are interested in design, they approach the field site with an open mind; they are open to the fieldwork changing their focus, project, or them. Elizabeth Meyer, in her well-titled and oft-cited chapter "Site Citations," points out the fallacy in the assumption that site analysis is rational. Meyer notes that "site concerns challenge the modern divide between rational site analysis and intuitive, creative conceptual design."[63] Indeed, site analysis cannot be entirely rational: How can we respond to a haptic, sensual entity with a purely rational approach? This holds true for garden design, as much as when we work in the realm of densely populated urban landscapes. While the elements of nature can be unruly, the most unpredictable element of nature is people. Potentially chaotic human behavior does not easily fit within rational quantifiable analysis—yet this is precisely why it is so important for landscape fieldworkers to try to understand it. The landscape fieldworker is interested not just in describing, but in prescribing and in the space between them.

Ethnography is often described as "thick description," a term coined by Gilbert Ryle and popularized by Clifford Geertz in his 1973 book, *The Interpretation of Cultures*.[64] Thick description is often used to describe an intellectual effort of ethnography where a wider interpretative context is described. While Geertz might consider that description is not prescriptive, landscape fieldworkers move between thick description and thick prescription, where the goal is to design projects that are deeply embedded in the operational fields of a landscape and effect wider spatial change. In gaining a thicker understanding of a landscape, action matters as much as reflection, and knowing when not to act.

Landscape is, to borrow from Michel de Certeau, a "practiced place."[65] Landscape fieldwork ultimately leads to the building of forms and interventions, books, and articles—and often through a circuitous, nonlinear route. That's because the design intentionality of fieldwork relies not just on what is found on the site but is also open to the unexpected, the haptic, and the imagination. Operational fields may come from the site as much as they may come from inside the landscape fieldworker. Landscape fieldwork is not done first, before a design process starts, and then set aside when the box is ticked. Landscape fieldwork continues throughout a design process and well after construction. Landscape fieldwork is integral throughout a design process.

At the height of her career, the anthropologist Margaret Mead was one of the most famous women in the world. Writing in the late 1960s about the lunar explorations, Mead took a surprising point of view. This was at a time when there were protests about the vast sums of money being spent in preparation for the voyage to the moon. Many anthropologists took the side of the subaltern, arguing for the money to be spent on poverty alleviation or population control, rather than space travel.[66] But Mead was adamant. Writing with her longtime collaborator and friend, Rhoda Métraux, they pointed out the projective capabilities of the human race: "Voyages to the moon—and beyond the moon—are one assurance of our ability to live on earth."[67] They argue that people collectively suffer from "future shock." "Tempocentrism," they argue, is an affliction worse than culture shock. They point out that humans suffer from two associated ailments: ethnocentrism and tempocentrism.[68] Ethnocentrism is when someone is excessively centered on one's own "'culture," and tempocentrism when someone is excessively centered on one's own timeframe. An example of tempocentrism might be environmental policy around climate change denial, for example.[69]

Creation of Knowledge

Aside from a few notable exceptions, landscape architecture has not developed the same set of skills for working with the social aspects of design as they have for the environmental and aesthetic ecologies. Part of the intention of this book has been to recenter people in discussions of landscape architecture, and to demonstrate that an understanding of social and cultural processes can lead to better foundations for a design project. Landscape fieldwork recognizes people as integral components of a landscape, equal partners with the plants, animals, soils and rocks, water and sky, objects and buildings, patches and corridors that comprise the environments we inhabit. If we understand landscape as the interaction of people and environment, then we must be able to trace and identify these human ecologies. To this end, the ethnographic naturally emerges as an essential tool to balance the geographic with the environmental and the aesthetic.

Decolonizing and Diversifying the Field

Despite an early fascination among the founders of the field with what the landscape anthropologist Tim Ingold describes as philosophizing out of doors,[70] landscape architecture has, to a large extent, forgotten fieldwork as an active part of a design research process. Out of doors, out of mind. There are of course notable exceptions in landscape architecture, and some of them referenced in this volume, but in general, we have forgotten its centrality, and how to talk about it. And, as I hope I have shown, landscape architects *do* do fieldwork; we just don't talk about it all that much. Landscape fieldwork today is often left to chance or intuition; students and junior landscape architects are sent out to the field with a notebook or iPhone in hand but little training on what to do, what to look at, how to take notes, never mind how to critique or interpret what they see. The huge potential for knowledge generation that is embedded in fieldwork lies forgotten and unactualized. As Albena Yaneva underscores, the slow and deliberate ethnographer is replaced with the hasty sightseer.[71] We can fix this by doing fieldwork.

As we've seen, landscape fieldwork is drawn from and learned from allied fields including anthropology, architecture, art, ecology, and urban planning.[72] Sometimes, landscape architecture has offered something back—such as with landscape and ecological urbanisms, which brought landscape architecture to architectural and other design audiences. But, by and large, the field is inward-looking, constrained by professional and disciplinary boundaries. This self-referentiality is something that will smother the discipline and stop it from taking on its rightful place as, as Jellicoe puts it, "the mother of all arts."[73] When landscape architecture does offer something back, it is usually within the established design fields. We should be more open to the skills of the landscape architect shaping the other arts and sciences too. Tackling this disciplinary inferiority complex head on should be a central concern of landscape architecture.

Furthermore, if we are to engage with cities, with governments, with the world, then landscape architects need to know how to deal with people, the most unpredictable and diverse element of the landscape. But this complex unpredictability—an essential aspect of our humanity—should be embraced rather than avoided. Part of the intention of this book has been to recenter people in the discussion of landscape architecture as active shapers of landscape along with plants, trees, soils, chemicals, and atmospheric processes. An awareness of the multiple ecologies—human, environmental, and aesthetic—leads to more successful understanding of landscape, and ultimately a more rigorous design process. These ecologies are not stable, they are moving and changing.

Forty years after Mead and Métraux, Bruno Latour wrote that we no longer have any certainties. Latour contrasts the lunar explorations of the 1960s with the space shuttle *Columbia* disaster of 2003. He tells us that "it is not only that time has passed. It has changed thoroughly from the way it used to pass."[74] Latour shakes our assumptions about modernity and leaves us with the images of the space shuttle *Columbia*'s take-off and photographs of the debris. We live in a world where past certainties are fragments, yet there is hope in the image too in terms of an infrastructural grid holding the pieces together.

> And yet of course everything is different from that age as well. Gaia is not the mere return of a pre-Galilean cosmos. Nothing much is left of modernism, and yet everything is new. To reengineer the imagination of architects, of urban planners (of Earth planners), of designers, of social scientists, of activists, and of citizens now unwittingly in charge of this planet, there is no way to turn to the past—there is no past to which to direct our queries, no help desk, no competent staff ready in Houston, and no gods or cherubs either. Those who have left the Cape Canaveral of modernism have to renew everything without the benefit of the modernist cosmos. They have to inhabit a wholly new place, a

place as different from a holy cosmos as it is from the cosmos of forty years ago. Where will they learn their new skills?[75]

Relatedly, if landscape architecture wants to be able to engage with the world in a more just way, it needs to be open to more diverse perspectives and voices. As a young discipline with a long history and a paucity of theory, landscape architecture has been around for only a relatively short time but claims a troubled lineage to the landscapes of the elites. The chosen histories of landscape fieldwork are often symbolic of power and exclusivity. Core precedents in landscape architecture, such as Blenheim, Rousham, not to mention Versailles, and even Shute, were built for elites. Some elitist landscapes, and emblems of the field, were built by the enslaved, or using funds generated from enslavement. One can think of the complicated history behind Kirstenbosch National Botanical Garden in Cape Town, for example: a stunning and botanically important landscape yet built by enslaved peoples.[76] Though this book has often looked to Jellicoe, Olmsted, and Burle Marx as inspiring examples, it is important to acknowledge that these origin myths of landscape architecture are grounded in certain white, male narratives. Jellicoe would likely be the first to admit that Lady Susan was a driving force behind his work. And Burle Marx that his chef, César, was something more than his chef.

If we were to envision a map of the world showing countries with professional associations of landscape architects, most of Africa is empty of such associations. Educators teaching about landscape on and in Africa refer to two main texts: *Cultural Landscape Heritage in Sub-Saharan Africa,* edited by John Beardsley and, more recently, *Landscapes: Canvas of Civilization* by Tunji Adejumo. Anthropologists have published much more about African landscape than have landscape architects.[77] If we are to create the knowledge that we need for diversifying the field of landscape architecture, we need to be paying close attention to local knowledge (i.e., the ways that people intervene in and use landscapes in daily life).

Landscape fieldwork can help to decolonize the field of landscape architecture. Landscape fieldwork leads us to unexpected outcomes because of this valuing of other ways of being. That is part of the point. It is—to use the overcited but nevertheless useful statement—about "making the familiar strange, and the strange

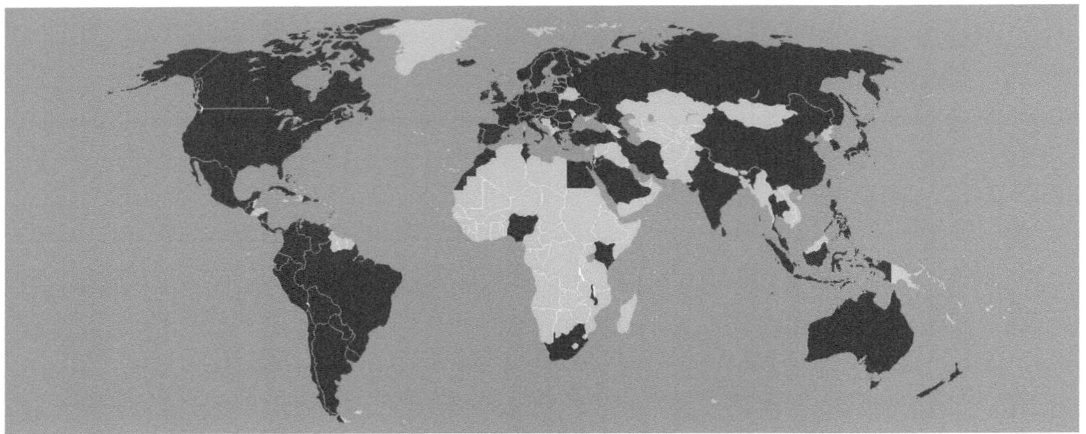

Figure 91. Countries with professional associations of landscape architects. Note that the profession of landscape architecture is largely absent in Africa. (Source: Mónica Pallares, International Federation of Landscape Architects [IFLA], 2023)

familiar."[78] This dialectical process is key: landscape fieldworkers must do a lot of reading to prepare for a project, including reading about the history of the site and its context, landscape architectural precedents, and methodology. At the same time, what we learn from fieldwork can contribute to broader theoretical understandings of landscape. Landscape fieldworkers do not necessarily do our philosophy out of doors, as Tim Ingold suggests,[79] but we do our philosophy out of the box.

To that point: This book began in Shute, a garden in the West of England. Fieldwork allowed me to learn about the importance of the garden's sacred spring, making it a sacred grove of sorts—a point that was left out of its design narrative. The Diamond was surrounded by a small grove of trees, until the Donegal County Council cut them down. The date palm groves of Bahrain had a similar fate, a fate that must be prevented for the Afro-Brazilian groves of Salvador da Bahia (chapter 5). In closing the book and considering these concerns with diversifying and decolonizing what landscape architecture might become, I'd like to move briefly to another landscape, in Nigeria. My fieldwork on Afro-Brazilian led me to one of the sources—sacred groves in West Africa. One of the best known extant sacred groves in Yorubaland is the Osun Sacred Grove in Osogbo, five or so hours' drive north of Lagos. The grove is home to many stories as well as being home for the orishas, including the deity Osun, who gracefully flow through this space of eighty or so hectares (about two hundred acres). The nearby Nelson Mandela Freedom Park, situated in the center of Osogbo, represents a more recent approach to the architecture of landscape. The park is described on the Osun State Government's website as a symbol of development that gives "the sleepy town a new look expected of a city."[80] This reference is a Western-centric one that might conjure images of New

Figure 92. Osogbo, Osun State, Nigeria. (Maria Vollas)

Figure 93. Arch of the Flying Tortoise, Osun Sacred Grove, designed by the artist and priestess Susanne Wenger after three weeks' meditation on the site. (Photograph by Adolphus Opara)

York's Central Park or London's Hyde Park. The park is as devoid of people as it is of trees. The Osun Sacred Grove and Nelson Mandela Freedom Park in Osogbo stand in stark contrast to each other.

But designers don't have to face a binary choice between the traditional and contemporary, indigenous and scientific, past versus future, culture and design.[81] We must avoid established hierarchies and recognize the indigenous design agency behind landscapes, especially where there is not a consolidated landscape architecture history or where such narratives are edited out of history. In doing so, we can

Figure 94. Osun Sacred Grove. (Photograph by Adolphus Opara)

draw from diverse sources and rich forms of knowledge that can help to nourish the profession globally. I am advocating not for traditional over contemporary, but for a recognition of the other as we critically imagine the future.[82]

This book has shown how landscape fieldwork can change the conception and boundaries of field sites, real and imagined. Inherent in all the projects is the belief that the landscape fieldworker has an intimacy with the site and its users. The examples advocate for forms of fieldwork that flatten social hierarchies, celebrate chance and serendipity, talk with inhabitants over extended periods of time, work collaboratively, use the whole body in the fieldwork process, and open up a creative space between description and prescription.

Epistemologies of landscape architecture—that is, how we know what we know—can only begin to be understood and enriched if we move outside of our self-referential worlds. By relying on Western, white, cisgender, mostly straight male histories alone, landscape architecture has eliminated large swaths of knowledge from their rightful place in landscape history. Landscape architecture was and is elitist, but since people inhabit landscapes, landscape architecture can affect the quality

Figure 95. Osun Sacred Grove. (Photograph by Adolphus Opara)

of life for all. No other discipline has the potential to engage humanity and other life forms in the ways that landscape architects can sometimes do. As a field concerned with designing the interrelationships of all living beings and in the process engaging all the human senses, from smell, to color, to touch, and taste, landscape architecture can offer paths to solutions to many of our most pressing problems, including addressing climate and social injustices. Yet, as a profession it continuously fails to be part of these urgent discussions in most of the world.

The Afro-Brazilian phrase *catar folhas* refers to a process of quiet work as "gathering leaves."[83] In the *terreiro,* as I learned, novices learn by working and doing. They

Figure 96. Osun Sacred Grove. (Photograph by Adolphus Opara)

Figure 97. Nelson Mandela Freedom Park, Osogbo, Nigeria. (Photograph by Adolphus Opara)

Figure 98. Osun Sacred Grove, Osogbo, Nigeria. (Photograph by Adolphus Opara)

don't ask questions—in fact, they're not encouraged to speak much. Instead, they learn through a process of osmosis, absorbing information that in due course adds up to an intergenerational transference of knowledge without texts. In the same way, in those part of the world devoid of landscape architects, and written landscape architectural histories, the landscape fieldworker must work patiently and collaboratively to accumulate the knowledge that can be documented and passed on to the profession in written form or other means.

Most of the world does not have the profession, discipline, or field of landscape architecture—but the entire world has landscape practices. Landscape fieldwork gives us the armature we need to rethink cohabitation on earth. Do we approach design with Western models that seem safe and proven to work for some privileged publics, or do we open design to new forms of knowledge to meet the massive challenges the whole world is facing? Innovation requires the humility to assume that landscape architecture—under its current formulation—is not always the most apt answer. The world fosters curiosity about the unknown, alterity, requiring openness to other ways of knowing. We can best learn such rich and radical forms of knowledge from the field through landscape fieldwork's direct, sustained, and embodied engagement.

GLOSSARY

Northern Irish–English Dialect (Ulster Scots)

Ballybeg	small town, from the Gaelic *Baile Beag*
brae	hillside or steep slope
glen	valley, from the Gaelic, *glean* (e.g., Glentogher)
kenter	market trader
lough	lake, or waterway, or arm of the sea, like the Scottish "loch"
mart	market (livestock)
Troubles	the 30-year period of armed conflict in Northern Ireland which ended with the Good Friday Peace Agreement of 1998
whin	furze, gorse, *Ulex europaeus*

Irish

Áras an Uachtaráin	residence of the president
baile	home, village, city (e.g., *abhaile, Baile Lifín, Baile Átha Cliath*)
clachan	small settlement
cnoc	hill (e.g., *Cnoc na Coille Dara,* the Hill of the Oak Wood).
cúil	woodland

Dáil Éireann	parliament of Ireland
Doire	Derry-Londonderry (meaning "Oak Grove")
glas	green
inis	island (e.g., Inis Eoghain, the island of Owen, transliterated as Inishowen)
láthar	center (of a settlement)
Teachta Dála	member of parliament
tírdhreach	landscape
sliabh	mountain (e.g., *Sliabh Sneacht,* meaning Snow Mountain)
straidbhaile	small settlement, literally "street town"
tuath	country; "*faoin tuath,*" literally "under the country"
Uachtarán	president

Arabic

adhān	call to prayer
Alachdar	*the Green*
Al-ḥamdu l-illāh	"Praise be to God"
'Ajam	Bahraini Sh'ia of Persian descent
'Ashura'	the anniversary of Imam Husayn, tenth day of Muharram
baḥr	*sea*
falaj (pl. aflāj)	water channel
ḥadith (pl. aḥādīth)	sayings of the Prophet Muhammad
Khaliji (noun, Khalij)	person from the Gulf, literally "Gulfie"
khuḍra	green / greenery
mashmūm	sweet basil
Muharram	first month of the Islamic calendar
sheikh (masc. plural shuyūkh, fem. sheikha, fem. pl. sheikhāt)	*member of ruling family, person with religious knowledge, or an elderly person, sheikh*
Shi'i (pl. Shi'a)	*follower of the Shi'a sect of Islam*
shlawnak?	literally, "What's your color?" ("How are you?")
suq	market
thawb (pl. thiyāb)	traditional men's garment, often white
qanā (pl. qanwāt)	underground water channel

Portuguese

abiã	a novice, "one who has kin by affinity"
arco-íris	rainbow
axé	the energy from which all life flows
babaorisha	Father of Saints
babalossain	Father of Leaves
Candomblé	Afro-Brazilian religion that fuses West African religious practice with elements of Christianity, in particular Catholicism
caminho	*way* (as in *Jesus é o caminho,* Jesus is the way)
cantam	sing
casa	house
catar folhas	gathering leaves
certo	sure!
ekedies	women who don't enter trance and who take care of the orishas
energia	energy
Exú	orisha of chance
folha	leaves
fundação	foundation (as in the Pierre Verger Foundation)
fuxicos	gossip
gringo	a nonlatino, especially from the United States
Iemenja	orisha of the sea, syncretized with the Virgin Mary (*Yemanjá*)
IPHAN	Brazilian National Institute of Historic and Artistic Heritage
Iroko	orisha of time
Iyakekerê	senior member of the community
iyalorixá	Mother of Saints (*mãe de santo*), abbreviated as iya (*iyaorishas*)
literatura de cordel	a form of popular literature associated with Northeastern Brazil
mãe de santo	Mother of Saints (*iyalorixá*)
motumbá	greeting (similar to *mo júbà,* "I pay homage")
motumbá axé	*reply to "motumbá"*
Nanã	orisha of mud
Obaluaye	orisha of earth and contagious disease
ogans	male members of the community who do not enter
Ogun	orisha of metals and war
orisha	a god, deity, or energy of nature (from the Yoruba, *òrìṣà*)

Ossain	orisha or energy of the leaves
Oxalá	the most senior of the orishas
Oxalufã	the older form of Oxalá
Oxóssi	orisha, or energy, of hunting, the forest, the arts and beautiful things.
Oxum	orisha of love and sweet waters
Oxumarê	orisha of the rainbow and the snake
Paulista	a person from São Paulo
rodante	a person with the gift of trance
sasanhe	singing leaves
terreiro	yards (*roças*) associated with Afro-Brazilian religions
Xangô	orisha of fire and justice
Yemanjá	orisha of the sea, syncretized with the Virgin Mary (Iemenja)

Yoruba

abiyan	*Candomblé novice*
àse	life energy
ìyákékeré	senior member of the community
ìyálòrìṣa	Mother of Saints
iyami	my mother
mo júbà	"I pay homage."
òrìṣà	orisha
ilé	home

Latin

mare	sea (e.g., Arab Mare)
id est viride	"That is green"
Sinus Persicus	Persian Sea

Latin (species)

Artocarpus heterophyllus	jackfruit
Crataegus monogyna	Common hawthorn
Ficus insipida	Insipid fig

Hedera hibernica	Irish ivy
Milicia excelsa	African teak
Terminalia catappa	Indian almond
Ulex europaeus	*furze, gorse, whin*

Greek

χώρα chora

Spanish

baja mar shallow water

NAMED PARTICIPANTS

Introduction

Lady Suzy Lewis	garden owner
Sir John Lewis	garden owner
Jala Makhzoumi	landscape architect
Günther Vogt	landscape architect

Chapter 1

John Bruton	taoiseach (prime minister of Ireland)
Margaret Butler	environmentalist
Anthony Cavanagh	construction worker
Faye Cunningham	chief planner
Michael Dander	construction manager
John Deaton	county architect
Máirín Nic Diarmada	intern
Eunan Doyle	quantity surveyor
Tom Farren	newspaper columnist
Richard Flynn	area engineer

Martha Freed	former department store owner
Brian Friel	playwright
Sir Geoffrey Jellicoe	landscape architect
Niall Kirkwood	professor of landscape architecture and technology, Harvard University
Paddy Harte	*Teachta Dála* (member of parliament)
Henry the Contractor	hardware merchant
John Hume	MP, MEP, Nobel Peace Prize winner
Pat Hume	John's wife
John Joe the Contractor	hardware merchant
Steven Izenour	architect
Rev. John Lappin	Presbyterian minister
Alexandra Lynch	office administrator
Tom McBrearty	retired headmaster
Tarla MacGabhann	architect
Joy McCormick	architect's wife
Liam McCormick	architect
Denis McGonagle	county councilor
Don McGonagle	project manager
Bernard McGuinness	county chairman
Eugene McGuinness	carpenter
Michael McLoone	county manager
Jim Melly	barman and undertaker
Ian L. McHarg	landscape architect
Ellen McIntyre	retired resident
Bernard Melly	ambulance driver
Danielle Moore	intern
Fr. Colm Morris	Catholic curate
Haruyoshi Ono	landscape architect
Patrick the Jeweler	jewelry store owner
Sean Potter	stone merchant
John Reid	landscape architect
Richard the Sailor	resident and shopowner
Mary Robinson	president of Ireland
Maud Smyth	elderly resident, mother of Beatrice and Iris
Anne Whiston Spirn	professor of landscape architecture and planning, MIT

Seamus Toland	paver
Milia Tsaoussis-Maddock	landscape architect

Chapter 2

Raoul Bunschoten	architect
James Corner	landscape architect
Takuro Hoshino	architect
Jeroen and his family members	residents

Chapter 3

Alireza	retired, Shiʿi (ʿAjam)
Aly	planner
Ahmed Kanna	anthropologist
crown prince	son of the king
Thomas Brandt Fibiger	anthropologist
Nelida Fuccaro	urban historian
Martin Hvidt	historian
Hussain Ilias	historian
Illias	taxi driver and handyman, Pakistani-born, Sunni
Isa	civil servant, Shiʿi (ʿAjam) and Sunni (Hawala) mix
Umm Isa	Isa's mother, Sunni (Hawala)
Dr. Juma	planner
Sawsan Karimi	anthropologist
Nabil	real estate agent, Lebanese Sunni
James Onley	historian
Hashim Sarkis	architect, dean of the School of Architecture and Planning, MIT
Sunny	research assistant
Dr. Youssef	architect and planner, Egyptian Copt

Chapter 4

Alex	student from Long Island, New York
Nicolette Bethel	anthropology professor at the College of The Bahamas
Mr. Bevans	resident of Exuma

Siri Linn Brandsøy	visual anthropologist
Caroline	business owner in Exuma
Steven Caton	anthropology professor at Harvard University
Construction Battalions . . .	US Navy Seebees
deputy prime minister of The Bahamas	officeholder varies; at the time of research it was Philip "Brave" Davis
Suneeta Rani Gill	visual anthropologist
Padraic Kelly	engineer
Hoon Kim	graphic designer
Kirby	US Navy Seebee, US Naval Construction Force (NCF), United States Navy
Mohsen Mostafavi	principal investigator and then dean of the Harvard Graduate School of design
Paul Nakazawa	architect
prime minister of The Bahamas	officeholder varies; at the time of research it was Perry Gladstone Christie
Rob	research associate at Harvard GSD
Sadie	graduate student
Shahab	student from the Middle East
Sunny	research assistant and student
Belinda Tato	principal of Ecosistema Urbano
T.J. business owner in Exuma	
Jose Luis Vallejo	principal of Ecosistema Urbano

Chapter 5

Alice	Candomblé practitioner and tour guide
Fábio	Candomblé initiate
Iya	person from Iyaorisha, Mother of Saints
Iyakekerê	elder in Candomblé community
Pai João	elder in Candomblé community
Mateus	anthropologist and Candomblé initiate

Chapter 6

Jala Makhzoumi	landscape architect

NOTES

Preface

1. With noted exceptions such as Rem Koolhaas and Aldo Rossi. See for example, Aldo Rossi, *A Scientific Autobiography* (Cambridge, MA: MIT Press, 1981); Luigi Ghirri, the photographer, complemented his photographs by writing in the first person. See *Luigi Ghirri, The Complete Essays, 1973–1991* (London: MACK, 2017).

2. Albena Yaneva, *Crafting History: Archiving and the Quest for Architectural Legacy* (Ithaca, NY: Cornell University Press, 2020), 20–21.

3. Tim Ingold, "Lines, Drawing, the Human Condition," a conversation between Tim Ingold, Momoyo Kaijima, Andreas Kalpakci and Anh-linh Ngo, translated from the German in issue 238 of *ARCH+* (March 2020), *Drawing Matter,* October 2021, accessed July 28, 2023, https://drawingmatter.org/lines-drawings-the-human-condition/.

4. Ingold, "Lines, Drawing, the Human Condition."

5. See Christopher Crouch and Jane Pearce. *Doing Research in Design* (London; Berg, 2012), 33–52 (especially 44–47).

6. John Monaghan and Peter Just, *Social and Cultural Anthropology: A Very Short Introduction* (Oxford: Oxford University Press, 2000), 27.

7. James Clifford and George E. Marcus, *Writing Culture: The Poetics and Politics of Ethnography* (Berkeley: University of California Press, 1986).

8. Donna Haraway, "Situated Knowledges: The Science Question in Feminism and the Privilege of Partial Perspective," *Feminist Studies* 14.3 (Autumn 1988): 575–99 (quote from 592).

9. Kenny Cupers, Sophie Oldfield, Manuel Herz, Laura Nkula-Wenz, Emilio Distretti, and Myriam Perret, *What Is Critical Urbanism?: Urban Research as Pedagogy* (Zürich: Park Books, 2022) 197.

10. In "Eidetic Operations and New Landscapes," Corner argues for the generative role of imaging in making landscapes. He occasionally writes in the first person, enabling readers to closely follow his thinking and calling their attention to critical points that he wants to emphasize. See James Corner, "Eidetic Operations and New Landscapes," in *Recovering Landscape: Essays in Contemporary Landscape Architecture,* ed. James Corner, 153–69 (New York: Princeton Architectural Press, 1999).

11. See Lawrence Halprin, *The RSVP Cycles; Creative Processes in the Human Environment* (New York: G. Braziller, 1970).

12. Randy Hester uses "I," "me," and "my" to describe his own experience of landscape design practice. See Randolph T. Hester, *Design for Ecological Democracy* (Cambridge, MA: MIT Press, 2006).

13. Jackson frequently writes in the first person, inviting readers into the everyday American landscapes through his words and drawings. See John Brinckerhoff Jackson, *Landscape in Sight: Looking at America,* ed. Helen Lefkowitz Horowitz (New Haven, CT: Yale University Press, 1997). Through the first person, Jackson also comments on various issues concerning American landscapes. See John Brinckerhoff Jackson, "The Necessity for Ruins," in *The Necessity for Ruins, and Other Topics,* 89–102 (Amherst: University of Massachusetts Press, 1980).

14. Gertrude Jekyll, the English garden designer, wrote a series of books in the first-person. See Gertrude Jekyll, *Children and Gardens. London: Country Life,* 1908; Gertrude Jekyll, *Colour Schemes for the Flower Garden,* 3rd ed. (London: Country Life, 1914); Gertrude Jekyll, *Home and Garden: Notes and Thoughts, Practical and Critical, of a Worker in Both* (London: Longmans, Green, 1900); Gertrude Jekyll, *Wood and Garden: Notes and Thoughts, Practical and Critical, of a Working Amateur* (London: Longmans, Green, 1926).

15. Geoffrey Jellicoe presents his critical readings of the Gardens of Ditchley Park by writing in the first person. See Geoffrey Jellicoe, "Ronald Tree and the Gardens of Ditchley Park: The Human Face of History," *Garden History* 10.1 (Spring 1982): 80–91.

16. Parts of McHarg 's *Design with Nature* were written in the first person; Ian L. McHarg, *Design with Nature,* 1st ed. (Garden City, NY: Natural History Press, for American Museum of Natural History, 1969).

17. Makhzoumi shares her first-hand experience with landscape architectural education in the

Arab Middle East. See Jala Makhzoumi, "Reflections on Landscape and Landscape Architecture Education in the Arab Middle East," in *The Routledge Handbook of Landscape Architecture Education,* ed. Diedrich Bruns and Stefanie Hennecke, 303–14 (New York: Routledge, 2022). Makhzoumi also draws upon her personal experience to discuss how landscapes empower the expression of citizenship. See Jala Makhzoumi, "Beirut's Public Realm and the Discourse of Landscape Citizenships," in *Landscape Citizenships,* ed. Tim Waterman, Jane Wolff, and Ed Wall, 182–204 (London; New York: Routledge, 2021).

18. Olin's observations about English landscapes, for example, are accompanied by sketches drawn in the field, which help bring readers closer to his own situated experience. See Laurie Olin, *Across the Open Field: Essays Drawn from English Landscapes* (Philadelphia: University of Pennsylvania Press, 2000).

19. Olmsted and Vaux write of Central Park in the first person to reveal their personal perception of the place as well as their design thinking and planning processes. See Frederick Law Olmsted Sr. and Calvert Vaux. "Description of a Plan for the Improvement of the Central Park, 'Greensward,' 1858," in *Forty Years of Landscape Architecture: Central Park,* ed. Frederick Law Olmsted Jr. and Theodora Kimball, 214–33 (Cambridge: MIT Press, 1973). Olmsted also writes in the first person about his visits to places such as The People's Park at Birkenhead and The North American Phalanx in New Jersey, sharing his observations and expressing his own feelings about the places. See Frederick Law Olmsted, "The People's Park at Birkenhead (1851)," in *Frederick Law Olmsted: Essential Texts,* ed. Robert Twombly, 39–48 (New York: W. W. Norton, 2010); Frederick Law Olmsted, "The Phalanstery and the Phalansterians (1852)," in *Frederick Law Olmsted: Essential Texts,* ed. Robert Twombly, 49–57 (New York: W. W. Norton, 2010).

20. See, for example, Anne Whiston Spirn, *The Language of Landscape* (New Haven: Yale University Press, 1998).

21. For example, in the introduction to *Lifeboat,* Stilgoe recalls many stories about lifeboats that he heard and experienced since his childhood. See John R. Stilgoe, introduction to *Lifeboat: A History of Courage, Cleverness, and Survival at Sea,* 1–19 (Charlottesville: University of Virginia Press, 2003).

22. Wall, for example, introduces his landscape mapping approaches in the first person. See Ed Wall, "Working with Uncertainties: Living with Masterplanning at Elephant and Castle," in *Landscape Citizenships,* ed. Tim Waterman, Jane Wolff, and Ed Wall, 205–24 (London: Routledge, 2021); Ed Wall, "Incompleteness: Landscapes, Cartographies, Citizenships," *Landscape Research* 47.2 (2022): 179–94.

23. In a journal article, Waterman writes in the first person to walk readers through the process of employing imagination in design. See Tim Waterman, "Making Meaning: Utopian

Method for Minds, Bodies, and Media in Architectural Design," *Open Library of Humanities* 4.1 (2018): 1–26.

24. See the collection of interviews in the first person in Günther Vogt, *Landscape as a Cabinet of Curiosities: In Search of a Position,* ed. Rebecca Bornhauser and Thomas Kissling (Zürich: Lars Müller, 2015).

25. Anne Whiston Spirn, online discussion with author, June 26, 2024.

26. Anne Whiston Spirn, *The Language of Landscape,* 62.

27. Anne Whiston Spirn, "The Eye Is a Door," in *The Eye Is a Door: Landscape, Photography, and the Art of Discovery* (Nahant, MA: Wolf Tree Press, 2014), para. 6, Kindle, accessed June 28, 2024, https://theeyeisadoor.com.

Prologue

1. Brian Friel, *Making History* (New York: Samuel French, 1989), 39.

2. Michel de Certeau, *The Practice of Everyday Life* (Berkeley: University of California Press, 1984), 117.

3. Siobhan McEvoy-Levy, "Youth Spaces in Haunted Places: Placemaking for Peacebuilding in Theory and Practice," *International Journal of Peace Studies* 17.2 (2012): 1–32.

4. Ireland, Bunreacht na hÉireann, *Constitution of Ireland* (Dublin: Oifig an tSoláthair, 1937).

5. Damian Holmes, "VPUU Harare Khyalitsha, Cape Town South Africa, Tarna Klitzner Landscape Architects," World Landscape Architecture, March 2014, accessed July 30, 2023, https://worldlandscapearchitect.com/vpuu-harare-khyalitsha-cape-town-south-africa-tarna-klitzner-landscape-architects/.

6. Jala Makhzoumi, "Colonizing Mountain, Paving Sea: Neoliberal Politics and the Right to Landscape in Lebanon," in *The Right to Landscape: Contesting Landscape and Human Rights,* ed. Shelley Egoz, Jala Makhzoumi, and Gloria Pungetti (Florence: Taylor & Francis, 2011), 227–44 (quote from 229).

7. Jala Makhzoumi, "Re-Defining Landscape: Erbil's Urban-Rural Greenbelt," *Perspecta: The Yale Architectural Journal* 53 (2020): 14–31 (quote from 30).

Introduction

1. John Stilgoe, *Outside Lies Magic: Regaining History and Awareness in Everyday Places* (New York: Walker, 1998), 1.

2. Jala Makhzoumi, online interview with the author, March 7, 2021.

3. Günther Vogt, personal communication, 2021.

4. Robert K. Yin, *Case Study Research and Applications: Design and Methods,* 6th ed. (Thousand Oaks, CA: Sage, 2018), xxi.

5. John Brinckerhoff Jackson, "The Need of Being Versed in Country Things," *Landscape* 1.1 (Spring 1951): 5.

6. Clifford Geertz, *The Interpretation of Cultures: Selected Essays* (New York: Basic Books, 1973), 10.

7. Michael Spens, "Admirable Jellicoe," *Architectural Review* 156.1111 (1989): 85–92.

8. Michael Spens, *Gardens of the Mind; The Genius of Geoffrey Jellicoe* (Woodbridge, UK: Garden Art Press, 1992), 12.

9. Charles E. Beveridge, "Frederick Law Olmsted Sr.," Olmsted Network, June 22, 2023, accessed July 6, 2022, https://olmsted.org/frederick-law-olmsted-sr/.

10. Sally Festing, *Gertrude Jekyll* (London: Viking, 1991), cover.

11. As quoted in Christopher Reed, "Nearsighted and Visionary," review of *Gertrude Jekyll* by Sally Festing (New York: Viking, 1991), *New York Times,* May 24, 1992, sec. 7, p. 20, https://www.nytimes.com/1992/05/24/books/nearsighted-and-visionary.html.

12. Festing, *Gertrude Jekyll, xi.*

13. These sketchbooks are now housed at Surrey History Centre in Woking and the Goldaming Museum.

14. Gertrude Jekyll, *Wood and Garden: Notes and Thoughts, Practical and Critical, of a Working Amateur* (Godalming, UK: Longmans, Green, 1899; reprint Woodbridge, UK: Antique Collectors' Club, 1981).

15. Judith Tankard, "Where Flowers Bloom in the Sands," *Country Life,* March 12, 1998, 82–85.

16. Francis Jekyll, Edwin Landseer Lutyens, Agnes Jekyll, and Jonathan Cape, *Gertrude Jekyll: A Memoir* (London: Jonathan Cape, 1934).

17. I heard this in conversation from a friend of Roberto Burle Marx, who tells me they would find him sleeping in the plant nursery.

18. Michael Spens quotes Jellicoe saying of his clients, "You have to get inside their minds." Michael Spens, *Jellicoe at Shute* (London: Academy Editions, 1993), 27.

19. Geoffrey Jellicoe, *Gardens of the Mind,* television program, dir. and prod. Roger Last, aired on February 17, 1991, on BBC Two, at 46:03 mins.

20. Spens, *Jellicoe at Shute,* 41.

21. Fieldnotes, October 1995.

22. Spens, *Jellicoe at Shute,* 45.

23. Spens, *Jellicoe at Shute,* 25.

24. J. C. Shepherd and G. A. Jellicoe, *Italian Gardens of the Renaissance* 3rd ed. (1925; London: Academy Editions, 1986), 50.

25. Geoffrey Jellicoe and Susan Jellicoe, *The Landscape of Man: Shaping the Environment from Prehistory to the Present Day* (London: Thames and Hudson, 1975).

26. M. Elen Deming and Simon Swaffield, *Landscape Architecture Research* (Hoboken, NJ: John Wiley & Sons, 2011), 153.

27. Spens, *Jellicoe at Shute,* 91.

28. As quoted in Hal Moggridge, "Jellicoe, Sir Geoffrey Alan (1900–1996), landscape architect," *Oxford Dictionary of National Biography,* September 23, 2004 (accessed September 24, 2022), https://www-oxforddnb-com.ezp-prod1.hul.harvard.edu/display/10.1093/ref:odnb/9780198614128.001.0001/odnb-9780198614128-e-40519.

29. Jellicoe, *Gardens of the Mind,* at 48:08 mins.

30. Stilgoe, *Outside Lies Magic,* 1.

31. Tim Ingold, "Anthropology Is *Not* Ethnography," *Proceedings of the British Academy* 154 (2008): 69–92 (quote from 83).

32. See Doherty, *Paradoxes of Green,* 26. See also, the example of the "Map of the World after the Destruction of the Tower of Babel" in chapter 3 of this book (page 00>).

33. Tim Ingold, *Making: Anthropology, Archaeology, Art and Architecture* (Abingdon, UK: Routledge, 2013).

34. This is the definition given on the website of the Department of Anthropology at Harvard University, accessed April 1, 2020, https://anthropology.fas.harvard.edu/.

35. Laurie Olin, online interview with the author, February 12, 2021.

36. Roger Sanjek, "A Vocabulary for Fieldnotes," in *Fieldnotes—The Makings of Anthropology,* ed. Roger Sanjek (Ithaca, NY: Cornell University Press, 1990), 92–121.

37. A common theme among anthropologists such as Andrea Ballestero, Eduardo Viveiros de Castro, Philippe Descola, and Eduardo Kohn.

38. "Archives and Community History. Wiltshire Council," accessed June 24, 2024, https://www.wiltshire.gov.uk/article/891/Archives-and-community-history.

39. Marilyn Strathern, *Commons and Borderlands: Working Papers on Interdisciplinarity, Accountability and the Flow of Knowledge* (Wantage, UK: Sean Kingston, 2004), 5–6.

40. In the United States, archaeology is technically considered part of anthropology. It's one of the four fields: archaeology, and biological/physical, linguistic, and cultural anthropology.

41. See, for example, Anne Elizabeth Yentsch, "Introduction: Close Attention to Place—Landscape Studies by Historical Archaeologists," in *Landscape Archaeology: Reading and*

Interpreting the American Historical Landscape, ed. Rebecca Yamin (Knoxville: University of Tennessee Press, 1996), xxiii–xlii.

42. "Archives and Community History. Wiltshire Council," accessed June 24, 2024, https://www.wiltshire.gov.uk/article/891/Archives-and-community-history.

43. Wiltshire Historic Buildings Trust, https://whbt.org.uk, accessed, September 23, 2022.

44. Doherty, "Is Landscape Literature?," in *Is Landscape . . . ? Essays on the Identity of Landscape, ed.* Gareth Doherty and Charles Waldheim (Abingdon, UK: Routledge, 2015), 13–43.

45. Ross Merrill, "Advice on Plein Air Equipment," *American Artist* 73.802 (2009): 28.

46. David Hockney and Martin Gayford, *A History of Pictures* (New York: Abrams, 2020), 208.

47. See "10 Things Aldo Leopold Used in the Field," on web page of *Wisconsin Magazine,* accessed, July 29, 2023, https://onwisconsin.uwalumni.com/10-things-aldo-leopold-used-in-the-field/.

48. Steven Handel, online discussion with the author, July 1, 2022.

49. Tankard, "Where Flowers Bloom in the Sands."

50. Yi Fu Tuan, foreword to *Geography and the Human Spirit,* by Anne Buttimer, ix–xi (Baltimore, MD: Johns Hopkins Press, 1993).

51. "Geography," National Geographic website, Education page, accessed, July 2, 2021, https://www.nationalgeographic.org/encyclopedia/geography/.

52. David Job, *New Directions in Geographical Fieldwork* (Cambridge: Cambridge University Press, 1999), 50.

53. Ann Blair, from her profile on the "People" page, Harvard Department of History website, accessed June 20, 2022, but no longer available, https://history.fas.harvard.edu/about.

54. John Dixon Hunt, "Is Landscape History?," in *Is Landscape . . . ? Essays on the Identity of Landscape,* ed. Gareth Doherty and Charles Waldheim (Abingdon, UK: Routledge, 2015), 247–60 (quote from 247).

55. Hunt, "Is Landscape History?," 247.

56. See Lynn Hunt, "How to Be an Historian: Five Tools," Los Angeles Community College website, accessed July 29, 2023, https://ilearn.laccd.edu/courses/85468/pages/how-to-be-an-historian-five-tools.

57. Shepherd and Jellicoe, *Italian Gardens of the Renaissance.*

58. From the seventeenth to the mid-nineteenth century, upper-class British gentlemen would engage in the Grand Tour, a form of art-historical fieldwork that usually brought them to Italy. There, they would experience ancient and Renaissance art, architecture, and gardens.

59. Jellicoe, *Gardens of the Mind,* at 20:05 mins.

60. Wiltshire and Swindon History Centre website, accessed July 29, 2023, https://wshc.org.uk/.

61. Jane Freeman and Janet H. Stevenson, "Parishes: Donhead St Mary," in *A History of the County of Wiltshire, vol. 13, South-West Wiltshire: Chalke and Dunworth Hundreds,* ed. D. A. Crowley (London: Victoria County History, 1987), 138–55. Digital version at British History Online, accessed May 7, 2021, http://www.british-history.ac.uk/vch/wilts/vol13/pp138-155.

62. See Museum of English Rural Life website, accessed June 24, 2024, https://merl.reading.ac.uk/collections/jellicoe-geoffrey/

63. See Garden Museum website, accessed July 29, 2023, https://gardenmuseum.org.uk/.

64. See "Library Catalogues," Lambeth Palace Library website, accessed July 29, 2023, https://www.lambethpalacelibrary.info/collections/library-catalogues/.

65. Frederick R. Steiner, *Making Plans: How to Engage with Landscape, Design, and the Urban Environment* (Austin: University of Texas Press, 2018), 1.

66. See William Hollingsworth Whyte, *The Social Life of Small Urban Spaces* (Washington, DC: Conservation Foundation, 1980).

67. Rahul Mehrotra, *Architecture in India—Since 1990* (Mumbai: Hatje Cantz, 2011), 57.

68. Steiner, *Making Plans.*

69. "Planning Policy," Wiltshire Council website, accessed July 29, 2023, https://www.wiltshire.gov.uk/planning-policy.

70. Wiltshire Council website, accessed June 24, 2024, https://www.wiltshire.gov.uk/article/854/Planning-and-building-control.

71. Joseph Ragsdale, "Fieldwork Hybrids: Learning from Other Disciplines How to Read, Record, and Reveal the Landscape," *Landscape Research Record* 7 (Blacksburg, VA: CELA Conference, 2018), 2.

72. Elizabeth Meyer, "Site Citations: The Grounds of Modern Landscape Architecture," in *Site Matters: Design Concepts, Histories, and Strategies, ed.* Carol J. Burns and Andrea Kahn (New York: Taylor & Francis, 2005, reprint 2021), 95–129, quote at 121.

73. Burns and Kahn, *Site Matters,* 3.

74. Marilyn Strathern, *Commons and Borderlands: Working Papers on Interdisciplinarity, Accountability and the Flow of Knowledge* (Wantage, UK: Sean Kingston, 2004), 5–6.

75. Strathern, *Commons and Borderlands,* 6.

76. Günther Vogt, online interview with the author, January 26, 2021.

77. For more on landscape meanings, see Gareth Doherty and Charles Waldheim, eds., *Is Landscape . . . ? Essays on the Identity of Landscape* (Abingdon, UK: Routledge, 2015); and

Gareth Doherty and Charles Waldheim, eds., *Landscape Is . . . ! Essays on the Meaning of Landscape* (Abingdon, UK: Routledge, forthcoming, 2025).

78. Gareth Doherty, "Is Landscape Literature?," in Doherty and Waldheim, *Is Landscape . . . ?*

79. Translations are considered in chapters 1 (Irish), 3 (Arabic), and 5 (Portuguese and Yoruba).

80. Geoffrey Jellicoe, "A Table for Eight," in *Space for Living: Landscape Architecture and the Allied Arts and Professions,* ed. Sylvia Crowe (Amsterdam: Djambatan, 1961), 13–21.

81. As translated by Ikram Saidane-Hamdi from the official title of the Tunisian Association of Landscape Architects and Engineers (TALAE).

82. See, for example, Gareth Doherty, "In the West You Have Landscape, Here We Have . . . ," *Studies in the History of Gardens and Designed Landscapes* 34.3 (2014): 201–6.

83. Subhadra Mitra Channa, "Contextualizing Indian Feminist Scholarship within Local Patriarchy and Global Influences," keynote lecture at the conference "What about Women in the History of Anthropology?," Federal University of Bahia, July 25, 2018.

1. Designing at Home

1. Ballybeg, meaning "small town" is the fictional village in which the playwright Brian Friel based many of his plays. I use the pseudonym to protect the identity of named characters.

2. The restaurant where I sat sketching my ideas was housed in an impressive modernist edifice designed by Scott Tallon Walker located three miles (about 5 kilometers) south of Dublin city center. UCD's Belfield campus had been designed by a young Polish-born architect, Andrzej Wejchert, the result of an architectural competition in the 1960s. I was inspired by Wejchert's ability to achieve what he did at a young age. At the same time there was something sinister about this campus, as its physical space was designed to minimize the threats of student rebellion.

3. See Coillín Aodha, "Fenton the Hangman," The Schools' Collection, vol. 0164, p. 145, National Folklore Collection UCD (University College Dublin) website, accessed July 30, 2023, https://www.duchas.ie/en/cbes/4672089/4669189/4678674.

4. Brian Friel, *The Gentle Island* (London: Davis Poynter, 1973), 26.

5. See Kevin Whelan, "The Modern Landscape," *Atlas of the Irish Rural Landscape,* ed. F. H. A. Aalen, Kevin Whelan, and Matthew Stout, rev. ed. (Cork: Cork University Press, 2011), 67–103, esp. 79–83.

6. The Plantation of Ulster was the organized "planting" of the northern part of Ireland with English and Scottish "planters," during the reign of James I (1603–25).

7. "Kenters" are market traders.

8. Maghtochair (pseud. of Michael Harkin), *Inishowen: Its History, Traditions, & Antiquities Containing a Number of Original Documents, with Numerous Notes from the Annals of the Four Masters, and Other Sources* (Londonderry: The [Derry] Journal Office, 1867).

9. Seamus Heaney, "The Sense of the Past," *History Ireland 1.4* (1993): 33.

10. Françoise Henry, *Irish Art in the Early Christian Period, to 800 A.D.* (Ithaca, NY: Cornell University Press, 1965), 118.

11. Henry, *Irish Art,* 118.

12. Henry, *Irish Art,* 128.

13. Anon., personal communication, 1995 (a professor at University College Dublin, who has since retired).

14. Robert Venturi, Denise Scott Brown, and Steven Izenour, *Learning from Las Vegas: The Forgotten Symbolism of Architectural Form* (Cambridge, MA: MIT Press, 1977), 87.

15. Venturi, Brown, and Izenour, *Learning from Las Vegas.*

16. Roberto Burle Marx (ca. 1992), in "Lost Paradise: The Gardens of Roberto Burle Marx," *Omnibus,* produced by the British Broadcasting Corporation, 2013, at 51 mins.; the program presented a profile of Burle Marx.

17. I had a copy of William Howard Adams, *Roberto Burle Marx: The Unnatural Art of the Garden* (New York: Museum of Modern Art: 1991), as well as a VHS of "Lost Paradise."

18. Haruyoshi Ono, personal communication, August 1996. Haruyoshi Ono was Burle Marx's long-term collaborator and heir.

19. I later edited these as lectures as a book, *Roberto Burle Marx Lectures: Landscape as Art and Urbanism* (Zürich: Lars Müller, 2018, reprint 2020).

20. See Geoffrey Jellicoe, *Studies in Landscape Design, vol. 2* (London: Oxford University Press, 1966), pl. 6b between pp. 10 and 11.

21. Quoted in Herbert Read, *A Concise History of Modern Painting* (New York: Praeger, 1959), 182.

22. Geoffrey Jellicoe, "Ourselves and History," *Landscape Architecture Magazine 49.1* (Autumn 1958): 28–32.

23. Geoffrey Jellicoe, personal communication, January 21, 1994.

24. Ruth Benedict, *The Chrysanthemum and the Sword: Patterns of Japanese Culture* (Boston, MA: Houghton Mifflin, 1946). The book has been harshly criticized, in part because Benedict did not do any fieldwork in Japan, and her work was conscripted by the U.S. Office of War Information for military purposes. It is a complicated and problematic text—and not considered a good source of information for understanding Japanese people and society. Had Benedict done fieldwork in Japan, the shortcomings might have been fewer.

25. See Kenneth Frampton, "Towards a Critical Regionalism: Six Points for an Architecture of Resistance," in *The Anti-Aesthetic: Essays on Postmodern Culture,* ed. Hal Foster (New York: New Press, 2002), 16–30.

26. Essentially "the spirit of the place." For more on *genius loci,* see Christian Norberg-Schulz, *Genius Loci: Towards a Phenomenology of Architecture* (New York: Rizzoli, 1980).

27. See Mariza G. S. Peirano, "When Anthropology Is at Home: The Different Contexts of a Single Discipline," *Annual Review of Anthropology* 27.1 (1998): 105–28. Also, Marilyn Strathern, "The Limits of Auto-Anthropology," in *Anthropology at Home, ed.* Anthony Jackson (London: Tavistock Publications, 1987), 16–37.

28. See the discussion around "halfie anthropology" in Lila Abu-Lughoud, "Writing against Culture," in *Recapturing Anthropology: Working in the Present,* ed. Richard G. Fox (Santa Fe, NM: School of American Research Press, 1996), 137–62 (quote from 138).

29. Fine Gael were then in power in the national government in a Rainbow coalition with the Labour Party and Democratic Left, a cross-spectrum alliance.

30. From the French, *Liaison entre actions de développement de l'économie rurale,* LEADER's seven principles for local development are as follows: that it be (1) area-based and (2) bottom-up; that it use (3) public-private partnerships, (4) innovation, (5) integration, (6) networking, and (7) cooperation.

31. *Lachico* is a slang term for a fool.

32. Witold Rybczynski, *Last Harvest: How a Cornfield Became New Daleville: Real Estate Development in America from George Washington to the Builders of the Twenty-First Century, and Why We Live in Houses Anyway* (New York: Scribner, 2007).

33. Rybczynski, *Last Harvest,* 241.

34. Rybczynski, *Last Harvest,* 71–77, see also references to road widths on 62, 127, and 241.

35. See "AAA Statement on Ethnography and Institutional Review Boards," America Anthropology Association website, accessed September 1, 2023, https://americananthro.org/about/policies/statement-on-ethnography-and-institutional-review-boards/. On the same page, the AAA goes on to explain: "Specifically, ethnography refers to the description of cultural systems or an aspect of culture based on fieldwork in which the investigator is immersed in the ongoing everyday activities of the designated community for the purpose of describing the social context, relationships and processes relevant to the topic under consideration."

36. The European Union started providing funding toward the peace process in 1989. For more on PEACE funding see "Northern Ireland PEACE PLUS Programme," European Parliament website, accessed July 30, 2023, https://www.europarl.europa.eu/factsheets/en/sheet/102/northern-ireland-peace-programme.

37. The EU provided €500 million and the British and Irish governments €167 million, according to Kenneth Bush and Kenneth Houston, *The Story of PEACE: Learning from European PEACE Funding in Northern Ireland and the Border Region* (Derry: INCORE, University of Ulster, 2011), 78.

38. "John Hume: Facts," The Nobel Prize website, accessed September 1, 2023, https://www.nobelprize.org/prizes/peace/1998/hume/facts/,

39. See "Practice: Stages of a Landscape Architecture Design Project," WLA website, accessed September 1, 2023, https://worldlandscapearchitect.com/practice-stages-of-a-landscape-architecture-design-project/.

40. See Brian A. Curran et al., *Obelisk: A History* (Cambridge, MA: MIT Press, 2009).

41. See A. T. Lucas, *Furze: A Survey and History of Its Uses in Ireland* (Dublin: National Museum of Ireland, 1960).

42. Lucas, *Furze,* 182.

43. *Landscape Design: The Journal of the Landscape Institute,* no. 273, London, September 1998, front cover and p. 44.

44. The project won a merit award from the Associated Landscape Contractors of America.

45. Tidy Towns Competition judging criteria, "Judging," SuperValu TidyTowns website, accessed July 30, 2023, https://www.tidytowns.ie/competition/judging/.

46. Tuesday, July 29, 2008.

47. Gabriel Arboleda, *Sustainability and Privilege: A Critique of Social Design Practice* (Charlottesville: University of Virginia Press, 2022), 135.

48. Tim Ingold, "The Temporality of the Landscape," *World Archaeology* 25.2 (1993): 152–74 (quote from 152–53).

49. Jan Gehl, *Cities for People* (Washington, DC: Island Press, 2010), 198.

50. Diana Wiesner, online interview with the author, March 29, 2021.

51. Tarla MacGabhann, online interview with the author, February 10, 2021.

52. MacGabhann, interview with the author, 2021.

53. MacGabhann, interview with the author, 2021.

2. Throwing Beans

1. Perhaps because of its in-betweenness, chora, Khôra, or χώρα has inspired many others too. Plato's "Timaeus" inspired Peter Eisenman and Jacques Derrida when they were collaborating on a garden for Parc de la Villette for instance. For more on *chora,* see John Sallis, *Chorology: On Beginning in Plato's Timaeus* (Bloomington: Indiana University Press, 2020).

2. Raoul Bunschoten and CHORA, *Urban Flotsam* (Rotterdam: 010 Publishers, 2001), n.p.

3. Bunschoten and CHORA, *Urban Flotsam,* unnumbered page.

4. Bunschoten and CHORA, *Urban Flotsam,* 206.

5. Raoul Bunschoten, D. Racz, and Alain J. Chiaradia, "Soul's Cycle," *A + U—Architecture and Urbanism* 263.8 (1992): 52–71 (quote from 52).

6. This raised a question of authorship, since the architect stands at a slightly ironic distance, as the one in control, the Master planner. The CHORA methodology can be seen as a critique of such an approach.

7. See Bunschoten, Racz, and Chiaradia, "Soul's Cycle."

8. Bunschoten and CHORA, *Urban Flotsam,* 203.

9. Bunschoten and CHORA, *Urban Flotsam,* 32.

10. Oxford English Dictionary online, accessed June 19, 2023.

11. Oxford English Dictionary online, accessed June 19, 2023.

12. Raoul Bunschoten and CHORA, "Urban Gallery," *ANC Architecture & Culture* 268 (Seoul, 2003): 87–99.

13. Oxford English Dictionary online, accessed June 19, 2023.

14. Raoul Bunschoten, "Stirring Still: The City Soul and Its Metaspaces," *Perspecta* 34 (2003): 56–65.

15. Bunschoten and CHORA, *Urban Flotsam.*

16. Bunschoten and CHORA, "Urban Gallery," 94–99.

17. See J. L. Borges, *A Universal History of Infamy* (London: Penguin Books, 1975).

18. Bunschoten and CHORA, *Urban Flotsam,* 209.

19. Bunschoten and CHORA, *Urban Flotsam,* 202.

20. Bunschoten and CHORA, "Urban Gallery," 94.

21. An example is illustrated in *Urban Flotsam* through a series of four photographs by Hélène Binet, Bunschoten's partner in life and one of the preeminent architectural photographers in the world; it shows the application of the four processes of EOTM to understand an everyday interaction in a coffee shop.

22. Bunschoten and CHORA, *Urban Flotsam,* 193.

23. Bunschoten and CHORA, "Urban Gallery," 94.

24. Bunschoten and CHORA, *Urban Flotsam,* 348; Bunschoten and CHORA, "Urban Gallery," 94.

25. Bunschoten and CHORA, *Urban Flotsam,* 359.

26. Bunschoten and CHORA, "Urban Gallery," 94–96.

27. Mohsen Mostafavi and Ciro Najle, "Prototype," in *Landscape Urbanism: A Manual for the*

Machinic Landscape, ed. Moshen Mostafavi and Ciro Najle (London: AA Publications, 2003), 89–101 (quote from 89).?

28. James Corner, "Not Unlike Life Itself: Landscape Strategy Now," *Harvard Design Magazine* 21 (Fall/Winter 2004): 32–34 (quote from 34).

29. Bunschoten and CHORA, *Urban Flotsam,* 359; Bunschoten and CHORA, "Urban Gallery," 96.

30. One might ask why an architectural project is used to illustrate the point of the landscape as a catalyst: the architecture was the prototype for landscape architectural reactions.

31. Raoul Bunschoten provided this example, which is explained in Ed Bacon's book *The Design of Cities.*

32. The AKDN is part of the Aga Khan Trust for Culture.

33. It may be tempting to confuse the term "prototype" with a catalyst. Prototypes are sometimes less accurately referred to in the literature as "catalysts," "activators," and "generators." They differ in that a catalyst is "a substance which, when present in small amounts, increases the rate of a chemical reaction or process but which is chemically unchanged by the reaction." In this sense, a landscape catalyst would remain unchanged while it triggers further reactions. Prototypes are something more. Prototypes do change and vary their composition over time. They are themselves fluid entities.

34. Kenneth Frampton, "Toward an Urban Landscape," *D: Columbia Documents of Architecture and Theory,* vol. 4 (1995), 83–93 (quote from 91).

35. Charles Waldheim, introductioni to *CASE Hilberseimer/Mies van Der Rohe, Lafayette Park Detroit* (Munich: Prestel; Cambridge, MA: Harvard University, Graduate School of Design, 2004), 19–24.

36. See Gareth Doherty, Stephen Ramos, and Hashim Sarkis, "*Dubai: Puerto como Prototipo,*" *Neutra* (2006): 26–30 (in Spanish, trans. Stephen Ramos).

37. See Rodolfo Machado, *The Favela-Bairro Project: Jorge Mario Jáuregui Architects* (Cambridge, MA: Harvard University Graduate School of Design, 2008). See also John Beardsley and Christian Werthmann, "Improving Informal Settlements: Ideas from Latin America," *Harvard Design Magazine* 28 (Spring/Summer 2008): 31–34.

38. James Corner, "Landscape Urbanism," in *Landscape Urbanism: A Manual for the Machinic Landscape,* ed. Mohsen Mostafavi and Ciro Najle (London: Architectural Association, 2003), 58–63 (quote from 59).

39. From the author's notes of a lecture by Stan Allen at the School of Architecture, Princeton University, March 3, 2005.

40. See George Lakoff and Mark Johnson, *Metaphors We Live By* (Chicago, IL: University of Chicago Press, 1980).

41. See Gilles Deleuze and Félix Guattari. *A Thousand Plateaus: Capitalism and Schizophrenia* (London: Athlone Press, 1988).

42. David Harvey, *Justice, Nature, and the Geography of Difference* (Cambridge, MA: Blackwell, 1996), as cited in James Corner, "The Agency of Mapping," in *Mappings,* ed. Denis E. Cosgrove (London: Reaktion Books, 1999), 214–53.

43. Keller Easterling, *Organization Space: Landscapes, Highways, and Houses in America* (Cambridge, MA: MIT Press, 1999), 2.

44. Corner, "Landscape Urbanism," 60.

45. From the author's notes of James Corner's keynote address to the symposium "Groundswell: Constructing the Contemporary Landscape" at the Cooper Union, New York, April 16, 2005.

46. Bunschoten, personal communication, 2022.

47. Günther Vogt, "Design Can Only Be Discussed between People on an Equal Footing," *Landscape as a Cabinet of Curiosities* (Zürich: Lars Müller, 2015), 91–128.

48. "Chora," database page, Spatial Agency website, https://www.spatialagency.net/database/chora.

49. Paulo Freire, *Pedagogy of the Oppressed, trans.* Myra Bergman Ramos (New York: Herder and Herder, 1970).

50. Augusto Boal, *Theater of the Oppressed,* (New York: Urizen Books, 1979). Thank you to Doris Sommer for her insights and introduction to Boal's work.

51. "Legislative Theatre," People Powered hub, accessed June 16, 2023, https://www.facebook.com/PeoplePoweredHub/posts/490970749287690/.

52. Literally, "string literature," *literatura de cordel* is a popular Latin American practice of displaying unbound booklets on string, like clothes on a clothes line. It is especially popular in the Brazilian northeast.

53. See "How" page, Pre-Texts website, accessed November 14, 2023, https://pre-texts.org/how/.

54. For more on pre-texts, see José Luis Falconi and Doris Sommer, eds., *Pre-Texts International* (Cambridge, MA: Harvard University Press, 2024).

55. Bunschoten and CHORA, *Urban Flotsam,* 359.

56. Bunschoten and CHORA, 2003, 98.

57. See Mihai Nadin, *Anticipation: The End Is Where We Start From* (Baden, Switzerland: Lars Müller, 2002), 35.

58. Niall G. Kirkwood, "Curating Resources," in *Ecological Urbanism,* ed. Mohsen Mostafavi and Gareth Doherty (Zürich: Lars Müller, 2016), 196–99.

59. Ekim Tan, "Negotiation and Design for the Self-Organizing City: Gaming as a Method for Urban Design," PhD diss., Delft University of Technology, 2014, 317.

60. Clifford Geertz, *The Interpretation of Cultures: Selected Essays* (New York: Basic Books, 1973), 5.

3. Walking and Talking

1. Gareth Doherty, *Paradoxes of Green: Landscapes of a City-State* (Oakland: University of California Press, 2017).

2. For more on the Bahraini dialect, see Clive Holes, *Colloquial Arabic of the Gulf and Saudi Arabia* (London: Routledge and Kegan Paul, 1984). My use of the terms "Persian Gulf" and "the Gulf" follows the majority international conventions; they have no political connotations.

3. Clifford Geertz, *The Interpretation of Cultures: Selected Essays* (New York: Basic Books, 1973), 21.

4. Roger Sanjek, "A Vocabulary for Fieldnotes," In *Fieldnotes—The Makings of Anthropology,* ed. Roger Sanjek (Ithaca, NY: Cornell University Press, 1990), 93.

5. For more on the temporal aspects of fieldwork, see Steven C. Caton's "ethnomemoir," *Yemen Chronicle: An Anthropology of War and Mediation* (New York: Hill and Wang, 2005).

6. Doherty, *Paradoxes of Green,* 4.

7. The *Oxford English Dictionary* dates the equation of green and environmentalism to West Germany in the early 1970s, and movements originating in campaigns against nuclear power stations such as the *Grüne Aktion Zukunft* (Green Campaign for the Future), and the *grüne Listen* (green lists) of election candidates. See the entry "green," in Oxford English Dictionary online, accessed July 24, 2023, http://www.oed.com/view/Entry/81167?rskey=Dnf5Kx&result=2#eid.

8. See "Guide to Country Comparisons," The World Factbook (CIA), accessed July 26, 2023, https://www.cia.gov/the-world-factbook/field/area/country-comparison/.

9. See "Climate Change Knowledge Portal," World Bank Group website, accessed July 26, 2023, https://climateknowledgeportal.worldbank.org/.

10. Ben Kendlim and Aubrey Herbert, writing of Bahrain in 1905, as quoted in Sir Charles Dalrymple Belgrave's autobiography, *Personal Column* (London: Hutchinson, 1960), 16.

11. "Bahrain population," accessed June 30, 2024, https://countrymeters.info/en/Bahrain. As of 2024, Bahrain's population is estimated at 1.8 million.

12. Residents of Ireland, the United Kingdom, and the United States have to this day special visa privileges.

13. As an area that is influenced by a particular city, of course Bahrain has a hinterland that is mostly in Saudi Arabia.

14. Albert J. Mills, Gabrielle Durepos, and Elden Wiebe, *Encyclopedia of Case Study Research* (Los Angeles, CA: Sage, 2010), 378–79.

15. Charles Waldheim, "Aerial Representation and the Recovery of Landscape," in *Recovering Landscape,* ed. James Corner (New York: Princeton Architectural Press, 1999), 136.

16. Richard T. T. Forman, *Land Mosaics: The Ecology of Landscapes and Regions* (Cambridge: Cambridge University Press, 1995), 3.

17. I am grateful to Samer al-Gilani and the staff at the Bayt al-Qur'an in Manama, who provided helpful assistance in searching the various books of the hadith for this phrase.

18. Introduction to this book, page XXX.

19. Bruce Jackson, *Fieldwork* (Urbana: University of Illinois Press, 1987), 19.

20. The program for the 2008 Gulf Studies conference (accessed, July 24, 2023) is at https://socialsciences.exeter.ac.uk/media/universityofexeter/instituteofarabandislamicstudies/images/gulf/2008_conference.pdf.

21. Nelida Fuccaro, *Histories of City and State in the Persian Gulf: Manama since 1800* (Cambridge: Cambridge University Press, 2009).

22. Charles Dalrymple Belgrave, *Personal Column* (London: Hutchinson, 1960).

23. By chance, I learned that the father of Joy McCormick, wife of the architect Liam McCormick mentioned in chapter 1, had known Charles Belgrave, and had visited their house when they were children.

24. These books include M. D. Cornes and C. D. Cornes. *The Wild Flowering Plants of Bahrain: An Illustrated Guide* (London: Immel, 1989); and Curtis E. Larsen, *Life and Land Use on the Bahrain Islands: The Geoarchaeology of an Ancient Society* (Chicago: University of Chicago Press, 1983).

25. These books include Haya Ali Al Khalifa and Michael Rice, eds., *Bahrain through the Ages: The History* (London: Routledge & Kegan Paul, 1986); and Shaikh Abdullah bin Khalid Al-Khalifa and Michael Rice, eds., *Bahrain through the Ages: The History* (New York: Routledge & Kegan Paul, 1993).

26. Nelida Fuccaro's 2009 urban history of Bahrain is an exemplary example. There are also more contemporary studies, such as the publications related to the various pavilions in the Venice Architecture Biennale. In 2010, Bahrain won the Golden Lion Prize for the best architecture

pavilion for the entry "Reclaim." See also Tarek Waly, *Private Skies: The Courtyard Pattern in the Architecture of the House* (Bahrain: Al Handasah Center, 1992).

27. For more on Patrick Abercrombie, see Mirjam Züger and Kees Christiaanse, eds., *The Potato Plan Collection: 40 Cities through the Lens of Patrick Abercrombie* (Rotterdam: Nai010 Publishers, 2018).

28. Apparently, A. M. Munro was influenced by Abercrombie, too. Educated in the United Kingdom, Dr. Youssef was at pains in our first meeting to point out that he was not a Muslim; he wanted me to know that he was a Christian, a Coptic Catholic at that. He had an accomplished career as an architect in Egypt and designed several Coptic churches. Now, among his many tasks, was advising the minister on the future of the Manama greenbelt, a topic that we discussed at great length over tea and biscuits in his office.

29. Yuri M. Lotman, "Text within a Text," *Soviet Psychology* 26.3 (Spring 1988): 32–51.

30. Albena Yaneva, *Crafting History: Archiving and the Quest for Architectural Legacy* (Ithaca, NY: Cornell University Press, 2020), 18.

31. Charles Waldheim, "Aerial Representation and the Recovery of Landscape," in *Recovering Landscape,* ed. James Corner (New York: Princeton Architectural Press, 1999), 136.

32. Walter Benjamin, *One-Way Street,* trans. Edmund Jephcott, ed. Michael W. Jennings (Cambridge, MA: Harvard University Press, 2016), 28–29.

33. Benjamin, *One-Way Street,* 28–29.

34. Francesco Careri, *Walkscapes: Walking as an Aesthetic Practice* (Barcelona: Editorial Gustavo Gili, 2002), back cover.

35. Doherty, *Paradoxes of Green,* 74.

36. Erving Goffman, "Regions and Region Behavior," in *The Presentation of Self in Everyday Life* (New York: Doubleday, 1990), 106–40 (quote from 112).

37. Giampietro Gobo and Andrea Molle, *Doing Ethnography* (New York: Sage, 2017), 288.

38. See preface to this volume.

39. Rereading previous notes every day is a practice I have heard attributed to Margaret Mead, as one of the reasons her ethnographic writing is so nuanced, although I have not found this reference in the literature.

40. Many anthropologists have written on the anthropologist as writer. See, for example: James Clifford and George E. Marcus, *Writing Culture: The Poetics and Politics of Ethnography* (Berkeley: University of California Press, 1986); and Clifford Geertz, *Works and Lives: The Anthropologist as Author* (Stanford, CA: Stanford University Press, 1988).

41. Galen Cranz, *Ethnography for Designers.* (Abingdon, UK: Routledge, 2016), 62, 63.

42. Josef Albers, *Interaction of Color* (New Haven, CT: Yale University Press, 1975), 1.

43. Doherty, *Paradoxes of Green,* 62–63.

44. James Corner, "The Agency of Mapping," in *Mappings,* ed. Denis E. Cosgrove (London: Reaktion Books, 1999), 230.

45. Gobo and Molle, *Doing Ethnography,* 287.

46. For more on embodied engagement, see Subhadra Mitra Channa, *The Inner and Outer Selves: Cosmology, Gender, and Ecology in the Himalayas* (Oxford: Oxford University Press, 2012).

47. A common description of the task of anthropologists is that they make "the familiar strange and the strange familiar." The poet Samuel Taylor Coleridge used this phrase to describe good poetry, as T. S. Eliot reports in his essay "Andrew Marvell," *Times Literary Supplement,* March 31, 1921.

48. While I have not found verification of this rule of thumb in the literature, I can attest to its accuracy from personal experience.

49. For more on ethnographing, see Gobo and Molle, *Doing Ethnography.*

50. Steven Caton, *Yemen Chronicle: An Anthropology of War and Mediation* (New York: Hill and Wang, 2005), 134.

51. Doherty, *Paradoxes of Green,* 53.

52. Fred Pearce, "Urban Heat: Can White Roofs Help Cool Earth's Warming Cities?" YaleEnvironment360 page, Yale School of the Environment, March 7. 2018, accessed July 26, 2023, https://e360.yale.edu/features/urban-heat-can-white-roofs-help-cool-the-worlds-warming-cities.

53. Pearce, "Urban Heat."

54. Doherty, *Paradoxes of Green,* 152–53.

55. Doherty, *Paradoxes of Green,* 51.

56. Doherty, *Paradoxes of Green,* 51.

57. Such as Andrea Ballestero, Dominic Boyer, George Marcus, Emily Sekine, and innovative journals such as *SAPIENS* (https://www.sapiens.org/).

4. Engaging Collectively

Eric Minns is a member of The Dicey Doh Singers, a group based in Nassau. To hear an extract from their "goombay" song quoted here and at the end of this chapter, see https://folkways.si.edu/dicey-doh-singers/you-aint-hurryin-me/world/music/track/smithsonian (accessed June 21, 2023).

1. Mohsen Mostafavi and Gareth Doherty, eds., *Ecological Urbanism* (Zürich: Lars Müller, 2010; rev. ed. 2016).

2. *Ecological Urbanism* was inspired by Guattari's "three ecologies": Félix Guattari, *The Three Ecologies,* trans. Ian Pindar and Paul Sutton (London: Athlone Press, 2000), 27–28.

3. The research project was sponsored by His Highness the Aga Khan as a gift to the Government of the Commonwealth of The Bahamas.

4. For more on the probable impacts of sea-level rise on The Bahamas, see Christina Gerhardt, *Sea Change: An Atlas of Islands in a Rising Ocean* (Berkeley: University of California Press, 2023).

5. For more on the political aspects of landscape, see Barbara Bender, ed., *Landscape: Politics and Perspectives* (Providence, RI: Berg, 1993), and Shelley Egoz, Jala Makhzoumi, and Gloria Pungetti, eds., *The Right to Landscape: Contesting Landscape and Human Rights* (London: Routledge, 2011).

6. The first public forum was held in Nassau on July 8, 2011, and the second on February 19, 2013.

7. The Jumentos Cays and Ragged Island Chain are located about one hundred miles (160 kilometers) southwest of Exuma and have a population of about seventy.

8. The exact site of Columbus's landing is disputed. Many assert it was Cat Island, known as San Salvador until 1925. Most maintain it was on the San Salvador formerly known as Watling's Cay.

9. For more on the color of water, see my chapter on blue: Gareth Doherty, "The Blueness of Green," in *Paradoxes of Green: Landscapes of a City-State* (Oakland: University of California Press, 2017), 44.

10. Edwin Ardener, "'Remote Areas'—Some Theoretical Considerations," in *The Voice of Prophesy and Other Essays* (New York: Berghahn Books, 1989, 2nd, expanded ed., 2018), 211–23. I thank Michael Herzfeld for introducing me to Ardener's work.

11. For more on human ecology, see Frederick B. Steiner, *Human Ecology: How Nature and Culture Shape Our World,* 2nd ed. (Washington, DC: Island Press, 2016).

12. Blue Cay, in the Exuma Cays Land and Sea Park, was advertised for $75 million in May 2020 by Sotheby's International Realty, website page discontinued, accessed May 1, 2020, https://www.sothebysrealty.com/eng/sales/detail/180-l-1145-7p2wrw/blue-island-exuma-cays-ex.

13. "Musha Cay and the Islands of Copperfield Bay," Venue Report website, accessed June 21, 2023, https://www.venuereport.com/venue/musha-cay-the-islands-of-copperfield-bay/. See also the Musha Cay website, http://www.mushacay.com/.

14. His Highness the Aga Khan generously sponsored the research project as a gift to the Bahamian nation.

15. See Bertin M. Louis Jr., *My Soul Is in Haiti: Protestantism in the Haitian Diaspora of the Bahamas* (New York: New York University Press, 2014).

16. "Immigration Fee Scale," Department of Immigration, Government of The Bahamas, accessed May 3, 2020https://www.immigration.gov.bs/applying-to-stay/immigration-fee-scale/.

17. See Anne Whiston Spirn, *The Language of Landscape* (New Haven: Yale University Press, 1998).

18. These drawn-out cycles can be difficult for ethnographers in other ways as well; PhD students must meet certain funding cycles for grant applications or find funding elsewhere, while early career scholars are left with little time to revise their own work as they compete for academic positions, not to mention the very lengthy processes of publishing articles and books.

19. Conrad M. Arensberg, *The Irish Countryman: An Anthropological Study* (1937; Prospect Heights, IL: Waveland Press, 1988).

20. Joel E. Savishinsky, *Strangers No More: Anthropological Studies of Cat Island, The Bahamas,* report of an ethnographic research project conducted in 1977 (Ithaca, NY: Ithaca College, Dept. of Anthropology, 1978).

21. Dominic Boyer and George E. Marcus, eds., *Collaborative Anthropology Today: A Collection of Exceptions* (Ithaca, NY: Cornell University Press, 2020).

22. A "starchitect" is a star architect, such as Rem Koolhaas, Thom Mayne, and the late Zaha Hadid.

23. I thank Michael Herzfeld for many conversations around this topic.

24. Timothy K. Choy et al., "A New Form of Collaboration in Cultural Anthropology: Matsutake Worlds," *American Ethnologist* 36.2 (2009): 380–403.

25. Nathaniel Samuel Murrell, "Obeah: Magical Art of Resistance," in *Afro-Caribbean Religions: An Introduction to Their Historical, Cultural, and Sacred Traditions* (Philadelphia, PA: Temple University Press, 2010), 231.

26. Ziying Tang, online interview with the author, January 15, 2021.

27. Clifford Geertz, *The Interpretation of Cultures* (New York: Basic Books, 1973), 6–7.

28. Morris P. Fiorina, "Extreme Voices: A Dark Side of Civic Engagement," in *Civic Engagement in American Democracy,* ed. Theda Skocpol and Morris P. Fiorina (Washington, DC: Brookings Institution Press, 2004), 395–425 (quote on 413).

29. The new government did still change the project's name, by insisting on the singular term "Exuma" for the archipelago, as opposed to "the Exumas," which is more commonly used in everyday speech. There may also be a slight geographic disparity between the terms: the Exumas being in the Exuma Cays, and Exuma being more centered on Great Exuma.

30. Mohsen Mostafavi, introduction to the Public Forum at the College of The Bahamas,

February 19, 2013, in *A Sustainable Future for Exuma: Environmental Management, Design, and Planning,* Engagement Report (Government of the College of The Bahamas, Bahamas National Trust, and Harvard Graduate School of Design, 2014), vol. 2, p. 53, available for download at http://www.sustainableexuma.org.

31. *The Tribune,* Nassau, Saturday, May 31, 2014, 13.

32. The coops are prototypes, in the sense of CHORA, the London-based office described in chapter 2.

33. Belinda Tato and Jose Luis Vallejo are principals of Ecosistema Urbano with offices in Madrid and Boston: https://ecosistemaurbano.com/.

34. The term "booxhibition" was later used for the *Atlas for a City-Region: Imagining the Post-Brexit Landscapes of the Irish Northwest.* Inspired by the Exuma project, the *Atlas for a City-Region* was based in the Critical Landscapes Design Lab at Harvard Graduate School of Design. It was sponsored by Derry City and Strabane District Council in Northern Ireland and Donegal County Council in Ireland.

35. Robert B. Textor, "The Value of Anticipation Despite Its Fallibility," *Ecological Urbanism, ed.* Mohsen Mostafavi and Gareth Doherty (Zürich: Lars Müller, 2016), 120–23.

36. Textor, "The Value of Anticipation," 120–23.

37. Textor, "The Value of Anticipation," 120–23.

38. We were complementing Stephen J. Pavlidis's *The Exuma Guide: A Cruising Guide to the Exuma Cays* (Port Washington, WI: Seaworthy Publications, 2010).

39. Richard T. T. Forman, *Land Mosaics: The Ecology of Landscapes and Regions* (Cambridge: Cambridge University Press, 1995), 13.

40. Forman, *Land Mosaics,* 13.

41. Forman, *Land Mosaics,* 13.

42. Carl Steinitz, "From Project to Global: On Planning and Scale," *Landscape Research* 9.2 (2004): 117–27.

43. Steinitz, "From Project to Global," 118.

44. Steinitz, "From Project to Global," 119.

45. Margaret Mead, in transcript of an interview with Ian McHarg, *The House We Live In,* aired on WCAU-TV, November 10, 1960, transcript in The Architectural Archives of the University of Pennsylvania, Ian L. McHarg Collection.

46. The videos are available on the Sustainable Exuma project's Vimeo channel, accessed June 23, 2023, https://vimeo.com/sustainableexuma.

47. We offered scholarships for Bahamians to study at the Harvard Graduate School of Design, both for the Design Discovery program in the summer, and for degree programs. Two

Bahamian students obtained these scholarships. Then we taught the executive education courses in the Bahamas, which became an important component of engagement. We developed a series of briefing documents on key themes and topics.

48. Diana Wiesner, online interview with the author, March 29, 2021.

49. Jala Makhzoumi, online interview with the author, March 7, 2021.

5. Gathering Leaves

1. In this chapter, when I use "Afro-Brazilian" it is to refer to Candomblé communities specifically.

2. Many scholars of African religions, including Jacob K. Olupona, suggest that West African gods cannot be understood from a Western perspective. See, for example, the introduction to Jacob K. Olupona, *City of 201 Gods: Ilé-Ifè in Time, Space, and the Imagination* (Berkeley: University of California Press, 2011), 1–17.

3. Other Afro-diasporic religions include Santería, Vodou, and Umbanda.

4. For more on trance see Bettina E. Schmidt, *Spirits and Trance in Brazil: An Anthropology of Religious Experience* (London: Bloomsbury Advances in Religious Studies, 2017), esp. chap. 3, "Experiencing and Explaining Ecstatic Religions: Religious Experience Revisited"; Paul Christopher Johnson, *Secrets, Gossip, and Gods: The Transformation of Brazilian Candomblé* (New York: Oxford University Press, 2002); and Jim Wafer, *The Taste of Blood: Spirit Possession in Brazilian Candomblé* (Philadelphia: University of Pennsylvania Press, 1992).

5. Many studies refer to sentient knowledge of trees and forests, among them Eduardo Kohn, *How Forests Think: Toward an Anthropology beyond the Human* (Berkeley: University of California Press, 2013).

6. For example, on the first page of their foundational text *Field and Laboratory Methods for General Ecology,* James Brower and Jerold Zar state that no matter how quantitative the methods that are used in a research project, it will succeed or fail based on the initial intuition of the researcher; the initial intuition of the researcher is essential; James E. Brower and Jerrold H. Zar. *Field and Laboratory Methods for General Ecology,* 4th ed. (Dubuque, IA: W. C. Brown, 1997).

7. Lila Abu-Lughoud, "Writing against Culture," in *Recapturing Anthropology: Working in the Present,* ed. Richard G. Fox (Santa Fe, NM: School of American Research Press, 1996), 137–62 (quote from 138).

8. The word *terreiro* is used by various Afro-Brazilian religions, including Umbanda and Candomblé. Here, it is used in the context of yards (*roças*) associated with Candomblé houses.

9. Vilson Caetano de Sousa Júnior, *Corujebó: candomblé e polícia de costumes (1938–1976)* (Salvador da Bahia: EDUFBA, 2018).

10. Once associated with every town in Yorubaland, for example, the groves of West Africa are largely depleted. The Osun Sacred Grove in Osogbo, with its fantastical sculptures by Susanne Wenger and the New Sacred Art Movement, is a rare exception (described in more detail in chapter 6).

11. Samuel Lira Gordenstein, "Planting Axé in the City: Urban Terreiros and the Growth of Candomblé in Late Nineteenth-Century Salvador, Bahia, Brazil," *Journal of African Diaspora, Archaeology, and Heritage* 5 (2016): 71–101 (information from 75).

12. Roger Bastide, *Le candomblé de Bahia* (Paris; La Haye: Mouton, 1958), 53.

13. In the 2022 census, Salvador emerged as the fifth-largest city in Brazil, having been overtaken by Fortaleza, the capital of Ceará state, also in the Northeast.

14. The Centro de Estudos Afro-Orientais (Center for Afro-Oriental Studies) at the Federal University of Bahia (UFBA) has catalogued 1,155 *terreiros on their website,* accessed June 30, 2024, https://terreiros.ceao.ufba.br/mapa.

15. Note that the number of spaces of worship is not necessarily a reflection of the number of adherents.

16. The Brazil Lab at Princeton University estimate that of the 12 million enslaved peoples brought from Africa to the New World, almost half were taken to Brazil between 1540 and the 1860s. Brazil Lab, Princeton University, "*Racialized Frontiers: Slaves and Settlers in Modernizing Brazil,*" Princeton Institute for International and Regional Studies, Princeton University, n.d., accessed July 8, 2023, https://brazillab.princeton.edu/research/racialized_frontiers.

17. Terreiros are spaces where minoritarian liberalisms are respected. See Moisés Lino e Silva on spaces of alternative forms of liberalism: Moisés Lino e Silva, *Minoritarian Liberalism: A Travesti Life in a Brazilian Favela* (Chicago: University of Chicago Press, 2022).

18. Mattijs van de Port, "Candomblé in Pink, Green and Black: Re-Scripting the Afro-Brazilian Religious Heritage in the Public Sphere of Salvador, Bahia," *Social Anthropology* 13.1 (2005): 3–26 (see 12–14), https://doi.org/10.1017/S0964028204001077. Some of the orishas are considered gender-fluid. For more, see Claudenilson da Silva Dias, *Identidades trans* em candomblés: entre aceitações e rejeições* (Simões Filho, Bahia: Editora Devires, 2020); and Roberto Strongman, *Queering Black Atlantic Religions: Transcorporeality in Candomblé, Santería, and Vodou* (Durham, NC: Duke University Press, 2019).

19. As quoted in van de Port, "Candomblé in Pink, Green and Black," 16.

20. Van de Port, "Candomblé in Pink, Green and Black," 12–14. See also Mattijs van de Port, *Ecstatic Encounters: Bahian Candomblé and the Quest for the Really Real* (Amsterdam: Amsterdam University Press, 2011), available at http://www.jstor.org/stable/j.ctt46n286.

21. Thank you to Ali Asani for introducing me to the term "experiential knowledge."

22. See Albena Yaneva, *Made by the Office for Metropolitan Architecture: An Ethnography of Design* (Rotterdam: 010 Uitgeverij, 2009). OMA is an architectural firm with offices in various locations worldwide.

23. Albena Yaneva, "How to Study the Ecology of Practice," in *Five Ways to Make Architecture Political: An Introduction to the Politics of Design Practice* (London: Bloomsbury, 2018), 33–52 (quote from 36).

24. Yaneva, "How to Study the Ecology of Practice."

25. For instance, the example of Geoffrey Jellicoe at Shute in the introduction.

26. Perhaps, this is one of the reasons there is not an established field of "landscape architecture" in Brazil. This linguistic lacuna was a matter of great frustration to landscape architects such as Roberto Burle Marx, who would often be dismissed as "gardeners" since people did not understand what landscape architects do. The spatial and social aspects of landscape architecture remain unactualized because of the lack of a word to describe this.

27. As described to me by Michelle Jean de Castro during fieldwork in Salvador da Bahia, July 2019.

28. John R. Stilgoe writes about the poetics of steps, explaining how the expression "a flight of stairs" recalls flights of birds and angels. And, I wonder, of orishas too. See John R. Stilgoe, "Design Standards: Whose Meanings?," in *Regulating Place: Standards and the Shaping of Urban America,* ed. Eran Ben-Joseph and Terry S. Szold (New York: Routledge, 2005), 17–44 (quote from 19).

29. Pierre Fatumbi Verger, *Ewé, the Use of Plants in Yoruba Society* (São Paulo: Editora Schwarcz, 1995).

30. Verger, *Ewé,* provides formulae for different ailments and desires, called *ebó.* Verger also documents recipes for magic spells to cast on people. Often Candomblé is criticized and misunderstood by outsiders for this magic.

31. *Axé* (*àṣẹ*) is at the center of Yoruba descendent religions. Everything and everyone has *axé;* it is carefully nurtured.

32. See Gareth Doherty and Moisés Lino e Silva, "Sixteen Plus One," *Harvard Design Magazine* 44 (2017): 79.

33. Olupona, *City of 201 Gods.*

34. Ruth Landes, *City of Women* (Albuquerque: University of New Mexico Press, 1994).

35. Fieldnotes, August 2016.

36. Fieldnotes, August 2016.

37. Vilson Caetano de Sousa Júnior, "*Kosi Ewé, Kosi Orisa: a natureza como lugar de destaque nas religiões afro-brasileiras*" (No leaves, no orisha: nature as a place of refuge for Afro-Brazilian

religions), conference presentation at Harvard Graduate School of Design, Cambridge, MA, October 3, 2019.

38. Sousa Júnior, "*Kosi Ewe, Kosi Orixá.*"

39. Pierre Verger, *Notas sobre o culto aos orixás e voduns na Bahia de Todos os Santos, no Brasil, e na antiga costa dos escravos, na África* (São Paulo: EDUSP, 1999), 517.

40. Sousa Júnior, "*Kosi Ewe, Kosi Orixá.*"

41. Stella de Oxóssi, *O que as folhas cantam (para quem canta folha),* 2nd. ed. (Rio de Janeiro: Autorale, 2020).

42. Sousa Júnior, "*Kosi Ewe, Kosi Orixá.*"

43. An *ekedie* is a female helper who does not enter trance.

44. Even though hugged by Oxóssi, the energy I received was from my own orisha, Oxalá, and more specifically Oxalufã, the older form of Oxalá.

45. Mãe Stella de Oxóssi (writing as Maria Stella de Azevedo Santos), *Meu tempo é agora* (My time is now). Salvador da Bahia: Assembleia Legislativa do Estado da Bahia, 2010, 163.

46. Marcio Goldman, "How to Learn in an Afro-Brazilian Spirit Possession Religion: Ontology and Multiplicity in Candomblé," in *Learning Religion: Anthropological Approaches,* ed. Ramón Sarró and David Berliner (Oxford: Berghahn Books, 2007) 103–19.

47. Goldman, "How to Learn," 114.

48. Goldman, "How to Learn," 109.

49. Van der Port, *Ecstatic Encounters.*

50. Orhan Pamuk, *My Name Is Red,* trans. Erdag M. Göknar (New York: Faber & Faber, 2001), 186.

51. Goldman, "How to Learn," 114.

52. Tim Ingold, "The Temporality of the Landscape," *World Archaeology,* 25.2 (1993): 152–74 (quote from 152).

53. Ingold, "The Temporality of the Landscape," 152–53.

54. Ingold, "The Temporality of the Landscape," 158.

55. The project is entitled "Terrific *Terreiros,*" and the research team includes Caio Frederico e Silva, Thiago Montenegro Góes, Lucidio Avelino, Vilma Patricia Silva, and Gareth Doherty.

56. Gary R. Hilderbrand, "Varied Tree Shade for New Urban Pleasures," *Harvard Design Magazine* 31 (Fall/Winter 2009): 79–85.

57. Fábio Macedo Velame, "Arquiteturas dos Terreiros de Candomblé: Entre os Africanistas e Crioulistas." *Anais do Seminário Arquitetura Vernácula*/ Popular. Anais. Salvador (BA) Programa de Pós-Graduação em Arquitetura e Urbanismo da UFBA, 2021.

58. Roger Bastide, *Le candomblé de Bahia,* 53.

59. *Instituto do Patrimonio Historico e Artistico Nacional:* see the IPHAN website, accessed July 25, 2023, http://portal.iphan.gov.br/.

60. Fieldnotes, August 2017. Verified in online discussion with Fábio Velame, June 27, 2024.

61. This figure was provided by Fábio Velame in an online discussion with the author, June 27, 2024.

62. Thanks to Fábio Velame for making this point in an email exchange, February 17, 2024. Velame attributes the term to Jussara Rego. Rego's text "Territórios do candomblé: a desterritorialização dos terreiros na Região Metropilitana de Salvador, Bahia," does not mention the term specifically. Rego acknowledges Rafael Soares de Olivera's article, "*Redes e Territórios: uma investigação hermenêutica sobre a história do candomblé da barroquinha.*" See Jussara Rego, "Territórios do candomblé: a desterritorialização dos terreiros na Região Metropilitana de Salvador, Bahia," *GeoTextos* 2.2 (2006): 31–85; and Rafael Soares de Oliveira, "Redes e Territórios: uma investigação hermenêutica sobre a história do Candomblé da Berroquinha," *Culture Vozes* 4.94 (2001): 44–61.

63. Felipe Olivera, "Encontro com a natureza é o destaque do Parque São Bartolomeu" (Meeting with nature is a delight in Parque São Bartolomeu), in *A cor da cidade,* Genesisfreitas blog post, November 27, 2015, accessed July 25, 2023, https://jornalacordacidade.wordpress.com/2015/11/27/parque-sao-bartolomeu-localizada-no-coracao-do-suburbio-ferroviario-de-salvador-o-parque-e-uma-reserva-florestal-incrivel/.

64. Fieldnotes, January 2019.

65. See, for example, J. Lorand Matory, *Black Atlantic Religion: Tradition, Transnationalism, and Matriarchy in the Afro-Brazilian Candomblé* (Princeton, NJ: Princeton University Press, 2005); and Alejandro de la Fuente and George Reid Andrews, eds., *Afro-Latin American Studies: An Introduction* (Cambridge: Cambridge University Press, 2018).

66. The Afro-Latin American Studies Institute (ALARI) at Harvard University was founded in 2013. Headed by Alejandro de la Fuente, ALARI was established in 2013 as the first US-based research institution to focus on the history and culture of peoples of African descent in Latin America and the Caribbean.

67. Duarte Vaz, online interview with the author, April 13, 2021.

68. Vilma Patricia Santana Silva, "Guided by Cowries, Designed by Orishas: Respecting the Architecture of Candomblé Terreiros," presentation to the colloqium "Sacred Groves and Secret Parks: Orisha Landscapes in Brazil and West Africa," Graduate School of Design, Harvard University, October 3–4, 2019, https://www.gsd.harvard.edu/event/sacred-groves-secret-parks-orisha-landscapes-in-brazil-and-west-africa/.

6. Thick Prescription

1. Landscape Architectural Accreditation Board, Accreditation Standards for Professional Programs in Landscape Architecture (Washington, DC: American Society of Landscape Architects, 2021), 1.

2. See Gareth Doherty, *Paradoxes of Green: Landscapes of a City-State* (Oakland: University of California Press, 2017), 153.

3. I refer to these terms not as binaries but as being inextricably intertwined. Aiwha Ong, Lila Abu-Lughod, and many others have cautioned against "culture"; "nature," likewise, is understood as being socially constructed. See, for example, Philippe Descola, *The Ecology of Others,* trans. Geneviève Godbout and Benjamin P. Luley (Chicago, IL: Prickly Paradigm, 2013).

4. Anne Whiston Spirn, *The Granite Garden: Urban Nature and Human Design* (New York: Basic Books, 1984), 3–6.

5. Several contemporary book reviews of Ian L. McHarg's *Design with Nature* pointed out this omission, among them Michael Laurie, "Book Review: *Design with Nature,*" *Architectural Research and Teaching* 1.2 (1970): 59–61; and R. Burton Litton Jr. and Martin Krieger, "Book Review: *Design with Nature,*" *Journal of the American Institute of Planners,* 37.1 (1971): 47–58.

6. For *Ecological Urbanism,* which I edited with Mohsen Mostafavi, I created a matrix of evaluation for each of these criteria.

7. Roberto Burle Marx, "The Garden as a Form of Art," in *Roberto Burle Marx Lectures: Landscape as Art and Urbanism,* ed. Gareth Doherty (Zürich: Lars Müller, 2018; repr. 2020), 120.

8. Burle Marx, "The Garden as a Form of Art," 120.

9. Burle Marx, "The Garden as a Form of Art," 228.

10. Burle Marx, "The Garden as a Form of Art," 228.

11. Burle Marx, "The Garden as a Form of Art," 228.

12. Stig L. Andersson, online interview with the author, March 7, 2021.

13. KDI was founded at the Harvard Graduate School of Design in 2006, by six students: Arthur Adeya, Patrick Curran, Chelina Odbert, Ellen Schneider, Jennifer Toy, and Kotchakorn Voraakhom. Voraakhom proposed the name, Kounkuey, which comes from the Thai word, คุ้นเคย, meaning "to know intimately."

14. For more on KDI, see, for example, Chelina Odbert and Joseph Mulligan, "The Kibera Public Space Project: Participation, Integration, and Networked Change," in *Now Urbanism: The Future City Is Here,* ed. Jeffrey Hou, Benjamin Spencer, Thaïsa Way, Ken Yocom (Abingdon, UK: Routledge, 2015), 177–92.

15. Jungyoon Kim, email to author, October 9, 2023.

16. See for example, Christopher Crouch and Jane Pearce, *Doing Research in Design* (London: Berg, 2012), 33–52.

17. Félix Guattari, *The Three Ecologies,* trans. Ian Pindar and Paul Sutton (London: Athlone Press, 2000), 27–28.

18. Spirn, *The Eye Is a Door,* 1.

19. John Brinckerhoff Jackson, "The Word Itself," in *Discovering the Vernacular Landscape* (New Haven, CT: Yale University Press, 1986), 1–8.

20. Richard T. T. Forman, *Land Mosaics: The Ecology of Landscapes and Regions* (Cambridge: Cambridge University Press, 1995), 3.

21. Forman, *Land Mosaics,* 13.

22. Carl Steinitz, "From Project to Global: On Planning and Scale," *Landscape Research* 9.2 (2004): 117–27.

23. Even landscapes that exist only in the mind are spatial. We might not measure it in feet and inches, or centimeters and meters, but in words and images. The "Record of the Make-Do Garden" was written by Huang Zhouxing (1611–80) to describe the garden he could not afford to own. It's a garden that exists only in words. Still, it is described in spatial, topographical terms: "All around the area are lofty mountains and steep ranges which encircle and embrace it like a lotus city."

24. Michel de Certeau, *The Practice of Everyday Life* (Berkeley: University of California Press, 1984), 117.

25. De Certeau, *The Practice of Everyday Life,* 117.

26. I.e., both qualitative and quantitative.

27. James Corner and Alex S. MacLean, *Taking Measures across the American Landscape* (New Haven, CT: Yale University Press, 1996).

28. Anuradha Mathur and Dilip da Cunha, *Mississippi Floods: Designing a Shifting Landscape* (New Haven, CT: Yale University Press, 2001).

29. See Philippe Descola, *Beyond Nature and Culture* (Chicago, IL: Chicago University Press, 2013).

30. Subhadra Mitra Channa, "Contextualizing Indian Feminist Scholarship within Local Patriarchy and Global Influences," keynote lecture at the conference *What about Women in the History of Anthropology?* Federal University of Bahia, July 25, 2018.

31. Subhadra Mitra Channa, "Embodied engagement," email, May 19, 2020.

32. Channa, email, 2020.

33. Channa, email, 2020.

34. Stig L. Andersson, online interview with the author, March 7, 2021.

35. Anne Whiston Spirn, *The Language of Landscape* (New Haven, CT: Yale University Press, 1998), 81.

36. Daniel Miller, *How to Conduct an Ethnography During Social Isolation,* YouTube video, May 30, 2020, accessed January 11, 2021, https://www.youtube.com/watch?v=NSiTrYB-oso.

37. Geoffrey Jellicoe, *Geoffrey Jellicoe: The Studies of a Landscape Designer over 80 Years,* vol. 1 (Woodbridge, UK: Garden Art Press, 1993), 16.

38. Geoffrey Jellicoe, interview with the author, January 1994.

39. Denis Wood, *The Power of Maps* (New York: Guilford Press, 1992).

40. Yaneva, *Made by the Office for Metropolitan Architecture,* 76.

41. Ingold's remark comes from a conversation between Tim Ingold, Momoyo Kaijima, Andreas Kalpakci, and Anh-linh Ngo, first published, in German translation, in issue 238 of *ARCH+* (March 2020). The original English version of the text was published on the Drawing Matter website (accessed September 18, 2022), https://drawingmatter.org/lines-drawings-the-human-condition/.

42. Robin Evans, "Translations from Drawing to Building." *AA Files* 12 (Summer 1986): 3–18.

43. Tim Ingold, "Anthropology Is *Not* Ethnography," *Proceedings of the British Academy* 154 (2008): 81.

44. Interview with P. Sullivan, *Sunday Times Magazine,* 1991, as quoted in Michael Spens, *Jellicoe at Shute* (London: Academy Editions, 1993), 27.

45. Yuri M. Lotman, "Text within a Text," *Soviet Psychology* 26.3 (Spring 1988): 32–51.

46. Lotman, "Text within a Text."

47. In James Bohman, Jeffrey Flynn, and Robin Celikates, "Critical Theory," in *The Stanford Encyclopedia of Philosophy* (Summer 2020 ed.), ed. Edward N. Zalta, accessed October 22, 2023, https://plato.stanford.edu/archives/sum2020/entries/critical-theory/, *Stanford Encyclopedia of Philosophy,* quoting Max Horkheimer, *Critical Theory* (New York: Seabury Press, 1972; repr. Continuum: New York, 1982), 246.

48. Bohman, Flynn, and Celikates, "Critical Theory," introduction.

49. Donald A. Schön, "Design as a Reflective Conversation with the Situation," in *The Reflective Practitioner: How Professionals Think in Action* (New York: Basic Books, 1983), 76–104.

50. Jacky Bowring, *Landscape Architecture Criticism* (Abingdon, UK: Routledge, 2020), 1.

51. Bowring, *Landscape Architecture Criticism,* 1.

52. "Critical Thinking," *Stanford Encyclopedia of Philosophy, July 2018,* accessed October 22, 2023, https://plato.stanford.edu/entries/critical-thinking/.

53. Anita Berrizbeitia, "Criticism in the Age of Global Disruption," *Journal of Landscape Architecture* 13.3 (2018): 24–27.

54. Oscar Wilde, *The Critic as Artist* (New York: David Zwirner Books, 2019).

55. Marc Treib, "Landscape Architecture Criticism," in *Land Forum,* ed. Anita Berrizbeitia (Fall / Winter 1997): 9–10 (quote from 9).

56. James Grant, *The Critical Imagination* (Oxford: Oxford University Press, 2013), 175.

57. Peter G. Rowe, *Design Thinking* (Cambridge, MA: MIT Press, 1987), 1.

58. Herbert Simon, *The Sciences of the Artificial* (Cambridge, MA: MIT Press, 1969), 55.

59. Robert B. Textor, "The Value of Anticipation Despite Its Fallibility," in *Ecological Urbanism,* ed. Mohsen Mostafavi and Gareth Doherty (Zürich: Lars Müller, 2016), 120–23.

60. Textor, "The Value of Anticipation."

61. Textor, "The Value of Anticipation."

62. Edgar Pieterse, endorsement for Kenny Cupers et al., eds., *What Is Critical Urbanism?: Urban Research as Pedagogy* (Zürich: Park Books, 2022), back cover.

63. Elizabeth Meyer, "Site Citations: The Grounds of Modern Landscape Architecture," in *Site Matters: Design Concepts, Histories, and Strategies, ed.* C. Burns and A. Kahn (New York: Taylor & Francis, 2005, reprinted 2021), 95–129 (quote from 93).

64. Clifford Geertz, "Thick Description: Toward an Interpretative Theory of Culture," in *The Interpretation of Cultures: Selected Essays* (New York: Basic Books, 1973), 1–35.

65. De Certeau, *The Practice of Everyday Life,* 117.

66. Robert B. Textor, introduction to *The World Ahead: An Anthropologist Anticipates the Future, by* Margaret Mead, ed. and comm. Robert B. Textor (New York: Berghahn Books, 2005), 1–31 (quote from 19).

67. Margaret Mead and Rhoda Métraux, "Man on the Moon," in Mead, *The World Ahead,* 247–52 (quote from 252).

68. Textor, introduction to Mead, *The World Ahead,* 16–18.

69. Textor, introduction to Mead, *The World Ahead,* 18.

70. Ingold, "Anthropology is *Not* Ethnography."

71. Yaneva, "How to Study the Ecology of Practice."

72. For more on the plurality of landscapes' identities and meanings, see Doherty and Waldheim, *Is Landscape . . . ? Essays on the Identity of Landscape;* and *Landscape Is . . . ! Essays on the Meaning of Landscape.*

73. Jellicoe, *Gardens of the Mind,* at 10:35–11:33 mins.

74. Bruno Latour, "Forty Years Later—Back to a Sub-lunar Earth," in *Ecological Urbanism, rev. ed., ed.* Mohsen Mostafavi and Gareth Doherty (Zürich: Lars Müller, 2016), 132.

75. Latour, "Forty Years Later," 133.

76. Melanie Boehl and Phakamani m'Afrika Xaba, "Decolonising Kirstenbosch: Confronting

the violent past of South Africa's botanical gardens," *AR, The Architectural Review*, January 28, 2021, accessed August 15, 2023, https://www.architectural-review.com/essays/decolonising-kirstenbosch-confronting-the-violent-past-of-south-africas-botanical-gardens.

77. For example, Trevor Marchant, *The Masons of Djenné* (Bloomington: Indiana University Press, 2009); Victor Turner, *The Forest of Symbols: Aspects of Ndembu Ritual* (Ithaca, NY: Cornell University Press, 1970); David William Cohen and E. S. Atieno Odhiambo. *Siaya: The Historical Anthropology of an African Landscape* (London: Ohio University Press, 1989); and Michael Bollig and Olaf Bubenzer, *African Landscapes: Interdisciplinary Approaches* (New York: Springer, 2009); James Fairhead and Melissa Leach, eds., *Misreading the African Landscape: Society and Ecology in a Forest-Savanna Mosaic* (Cambridge: Cambridge University Press, 1996).

78. Coleridge, quoted in Eliot, "Andrew Marvell."

79. Ingold, "Anthropology is *Not* Ethnography."

80. Website of the State of Osun, 2021, https://www.osunstate.gov.ng/2021/.

81. Dilip M. Menon, *Changing Theory: Concepts from the Global South* (Milton Park, UK: Routledge, 2022).

82. See my article in the *African Journal of Landscape Architecture*, "Culture, Design, and Two Yoruba Landscapes," 2021, https://www.ajlajournal.org/articles/culture-design-and-two-yoruba-landscapes.

83. Marcio Goldman, "How to Learn in an Afro-Brazilian Spirit Possession Religion: Ontology and Multiplicity in Candomblé," in *Learning Religion: Anthropological Approaches*, ed. Ramón Sarró and David Berliner (Oxford: Berghahn Books, 2007), 103–19 (quote from 109); and see chapter 5 in this volume.

BIBLIOGRAPHY

Aalen, F. H. A., Kevin Whelan, and Matthew Stout. *Atlas of the Irish Rural Landscape.* Toronto: University of Toronto Press, 1997.

Abu-Lughoud, Lila. "Writing against Culture." In *Recapturing Anthropology: Working in the Present,* edited by Richard G. Fox, 137–62. Santa Fe, NM: School of American Research Press, 1996.

Adams, William Howard. *Roberto Burle Marx: The Unnatural Art of the Garden.* New York: Museum of Modern Art, 1991.

Albers, Josef. *Interaction of Color.* New Haven, CT: Yale University Press, 1975.

Al Khalifa, Shaikha Haya Ali, and Michael Rice, eds. *Bahrain through the Ages: The History.* London: Routledge & Kegan Paul, 1986.

Al-Khalifa, Shaikh Abdullah bin Khalid, and Michael Rice, eds. *Bahrain through the Ages: The History* (New York: Routledge & Kegan Paul, 1993).

Arboleda, Gabriel. *Sustainability and Privilege: A Critique of Social Design Practice.* Charlottesville: University of Virginia Press, 2022.

Ardener, Edwin. "'Remote Areas'—Some Theoretical Considerations." In *The Voice of Prophesy and Other Essays,* 211–23. New York: Berghahn Books, 1989; 2nd, expanded ed., 2018.

Arensberg, Conrad M. *The Irish Countryman: An Anthropological Study.* New York: Macmillan, 1937; Prospect Heights, IL: Waveland Press, 1988.

Bastide, Roger. *Le candomblé da Bahia.* Paris: Mouton, 1958.
Bateson, Gregory. *Steps to an Ecology of Mind: Collected Essays in Anthropology, Psychiatry, Evolution, and Epistemology.* New York: Ballantine Books, 1977.
Beardsley, John, and Christian Werthmann. "Improving Informal Settlements: Ideas from Latin America." *Harvard Design Magazine* 28 (Spring/Summer 2008): 31–34.
Belgrave, Charles Dalrymple. *Personal Column.* London: Hutchinson, 1960.
Bender, Barbara, ed. *Landscape: Politics and Perspectives.* Providence, RI: Berg, 1993.
Benedict, Ruth. *The Chrysanthemum and the Sword: Patterns of Japanese Culture.* Boston, MA: Houghton Mifflin, 1946.
Benjamin, Walter. *One-Way Street.* Translated by Edmund Jephcott, edited by Michael W. Jennings. Cambridge, MA: Harvard University Press, 2016.
Berlanda, Tomà, Meghan Ho-Tong, Marah Khalifeh, Adila Laïdi-Hanieh, eds. *Landwalks across Palestine and South Africa.* Cape Town: University of Cape Town and The Palestinian Museum, 2022.
Berrizbeitia, Anita. "Criticism in the Age of Global Disruption." *Journal of Landscape Architecture* 13.3 (2018): 24–27.
———. "Landscape Architecture Criticism." *Land Forum* (Fall/Winter1997): 2.
Blier, Suzanne Preston. *The Anatomy of Architecture: Ontology and Metaphor in Batammaliba Architectural Expression.* Cambridge: Cambridge University Press, 1987.
Boehl, Melanie, and Phakamani m'Afrika Xaba. "Decolonising Kirstenbosch: Confronting the Violent Past of South Africa's Botanical Gardens." *AR, The Architectural Review* 28 (January 2021), n.p. https://www.architectural-review.com/essays/decolonising-kirstenbosch-confronting-the-violent-past-of-south-africas-botanical-gardens.
Bowring, Jacky. *Landscape Architecture Criticism.* Abingdon, UK: Routledge, 2020.
Boyer, Dominic, and George E. Marcus, eds. *Collaborative Anthropology Today: A Collection of Exceptions.* Ithaca, NY: Cornell University Press, 2020.
Brower, James E., and Jerrold H. Zar. *Field and Laboratory Methods for General Ecology.* 4th ed. Dubuque, IA: W. C. Brown, 1997.
Bunschoten, Raoul. "The Architect as Curator," *Hunch* (2003): 120–22.
———. "CHORA." In *Architectural Design,* edited by Jane Anderson, 82–99. Lausanne: AVA Publishing, 2011.
———. "Points, Spirals and Prototypes." In *Architecture and Participation,* edited by Peter Blundell Jones, Doina Petrescu, and Jeremy Till. London: Routledge, 2005.
———. "Stirring Still: The City Soul and Its Metaspaces." *Perspecta,* 34 (2003): 56–65.

———. "Urban Prototypes." In *Ecological Urbanism, edited by Mohsen Mostafavi and Gareth Doherty,* 826–31. Zürich: Lars Müller, 2016.
Bunschoten, Raoul, and *CHORA. From Matter to Metaspace: Cave, Ground, Horizon, Wind.* New York: Springer, 2006.
———. *Public Spaces.* London: Black Dog Publishing, 2002.
———. "Stirring the City" and "Urban Gallery." *ANC Architecture & Culture* 268 (Seoul, 2003): 87–99.
———. *Urban Flotsam.* Rotterdam: 010 Publishers, 2001.
Bunschoten, Raoul, D. Racz, and Alain J. Chiaradia. "*Soul's Cycle.*" *A + U—Architecture and Urbanism* 263.8 (1992): 52–71.
Burns, Carol, and Andrea Kahn, eds. *Site Matters: Design Concepts, Histories, and Strategies.* New York: Taylor & Francis, 2005; reprinted 2021.
Bush, Kenneth, and Kenneth Houston. *The Story of PEACE: Learning from European PEACE Funding in Northern Ireland and the Border Region.* Derry: INCORE, University of Ulster, 2011.
Careri, Francesco. *Walkscapes: Walking as an Aesthetic Practice.* Barcelona: Editorial Gustavo Gili, 2002.
Caton, Steven. *Yemen Chronicle: An Anthropology of War and Mediation.* New York: Hill and Wang, 2005.
Certeau, Michel de. *The Practice of Everyday Life.* Berkeley: University of California Press, 1984.
Channa, Subhadra Mitra. *The Inner and Outer Selves: Cosmology, Gender, and Ecology in the Himalayas.* Oxford: Oxford University Press, 2012.
Clifford, James, and George E. Marcus. *Writing Culture: The Poetics and Politics of Ethnography.* Berkeley: University of California Press, 1986.
Corner, James. "Eidetic Operations and New Landscapes." In *Recovering Landscape: Essays in Contemporary Landscape Architecture,* edited by James Corner, 153–69. New York: Princeton Architectural Press, 1999.
———. "Landscape Urbanism." In *Landscape Urbanism. A Manual for the Machinic Landscape, edited by* Mohsen Mostafavi and Ciro Najle, 58–63. London: Architectural Association, 2003.
———. "Not Unlike Life Itself: Landscape Strategy Now." *Harvard Design Magazine* 21 (Fall/Winter 2004): 32–34.
———. "Representation and Landscape: Drawing and Making in the Landscape Medium." *Word & Image: A Journal of Verbal/Visual Enquiry* 8.3 (1992): 243–75.
———. "The Agency of Mapping." In *Mappings,* edited by Denis E. Cosgrove, 214–53. London: Reaktion Books, 1999.

Corner, James, and Alex S. MacLean, *Taking Measures across the American Landscape.* New Haven, CT: Yale University Press, 1996.

Cornes, M. D., and C. D. Cornes. *The Wild Flowering Plants of Bahrain: An Illustrated Guide.* London: Immel, 1989.

Cranz, Galen. *Ethnography for Designers.* Abingdon, UK: Routledge, 2016.

Crouch, Christopher, and Jane Pearce. *Doing Research in Design.* London: Berg, 2012.

Cupers, Kenny, Sophie Oldfield, Manuel Herz, Laura Nkula-Wenz, Emilio Distretti, and Myriam Perret. *What Is Critical Urbanism?: Urban Research as Pedagogy.* Zürich: Park Books, 2022.

Deleuze, Gilles, and Félix Guattari. *A Thousand Plateaus: Capitalism and Schizophrenia.* London: Athlone Press, 1988.

Deming, M. Elen, and Simon Swaffield. *Landscape Architecture Research.* Hoboken, NJ: John Wiley & Sons, 2011.

Descola, Philippe. *Beyond Nature and Culture.* Chicago: Chicago University Press, 2013.

———. *The Ecology of Others.* Translated by Geneviève Godbout and Benjamin P. Luley. Chicago: Prickly Paradigm, 2013.

Doherty, Gareth. "Culture, Design, and Two Yoruba Landscapes." *African Journal of Landscape Architecture* (2021), n.p. https://www.ajlajournal.org/articles/culture-design-and-two-yoruba-landscapes.

———. "Dubai's Lopsided Landscape." In *The Superlative City: Dubai and the Urban Condition in the Early 21st Century,* edited by Ahmed Kanna, 53–63. Cambridge, MA: Aga Khan Program, Harvard University Graduate School of Design, 2013).

———. "In the West You Have Landscape, Here We Have . . ." *Studies in the History of Gardens and Designed Landscapes* 34.3 (2014): 201–6.

———. *Paradoxes of Green: Landscapes of a City-State.* Oakland: University of California Press, 2017.

———, ed. *Roberto Burle Marx Lectures: Landscape as Art and Urbanism.* Zürich: Lars Müller, 2018.

Doherty, Gareth, and Moisés Lino e Silva. "Sixteen Plus One." *Harvard Design Magazine* 44, (2017): 79.

Doherty, Gareth, Stephen Ramos, and Hashim Sarkis. "Dubai: Puerto como Prototipo." *Neutra.* (2006): 26–30. In Spanish, translation by Stephen Ramos.

Doherty, Gareth, and Charles Waldheim, eds. *Landscape Is . . . ! Essays on the Meaning of Landscape.* Abingdon, UK: Routledge, forthcoming in 2025.

———. *Is Landscape . . . ? Essays on the Identity of Landscape.* Abingdon, UK: Routledge, 2016.

Don, Monty. "Stream of Subconsciousness." From *The Secret History of the British Garden.* Aired December 1, 2015, on BBC Two. https://www.bbc.co.uk/programmes/p039tpmx.

Duarte, Fábio, Gareth Doherty, and Paul Nakazawa. "Redrawing the Boundaries: Planning and Governance of a Marine Protected Area—The Case of the Exuma Cays Land and Sea Park." *Journal of Coastal Conservation* 21.2 (April 201): 265–71.

Easterling, Keller. *Organization Space: Landscapes, Highways, and Houses in America.* Cambridge, MA: MIT Press, 1999.

Egoz, Shelley, Jala Makhzoumi, and Gloria Pungetti, eds. *The Right to Landscape: Contesting Landscape and Human Rights.* London: Routledge, 2011.

Eliot, T. S. "Andrew Marvell." *Times Literary Supplement,* March 31, 1921.

Elsheshtawy, Yasser. "Sheltering Space: Little Bangladesh in Abu Dhabi." In *Temporary Cities: Resisting Transience in Arabia,* 134–69. London: Routledge, 2019.

Ewing, Suzanne, Jeremie Michael McGowan, Chris Speed, and Victoria Clare Bernie. *Architecture and Field/Work.* London: Routledge, 2011.

Falconi, José Luis, and Doris Sommer, eds. *Pre-Texts International.* Cambridge, MA: Harvard University Press, forthcoming 2024.

Festing, Sally. *Gertrude Jekyll.* London: Viking, 1991.

Fiorina, Morris P. "Extreme Voices: A Dark Side of Civic Engagement." In *Civic Engagement in American Democracy,* edited by Theda Skocpol and Morris P. Fiorina, 395–426. Washington, DC: Brookings Institution Press and Russell Sage Foundation, 1999.

Flyvbjerg, Bent. "Five Misunderstandings about Case-Study Research." *Qualitative Inquiry* 12.2 (2006): 219–45.

Forman, Richard T. T. *Land Mosaics: The Ecology of Landscapes and Regions.* Cambridge: Cambridge University Press, 1995.

Frampton, Kenneth. "Toward an Urban Landscape." In *D: Columbia Documents of Architecture and Theory, vol. 4,* 83–93. New York: Columbia University, 1995.

———. "Towards a Critical Regionalism: Six Points for an Architecture of Resistance." In *The Anti-Aesthetic: Essays on Postmodern Culture, edited by Hal* Foster, 16–30. New York: New Press, 2002.

Freire, Paulo. *Pedagogy of the Oppressed.* Translated by Myra Bergman Ramos. New York: Herder and Herder, 1970

Friel, Brian. *The Gentle Island.* London: Davis Poynter, 1973.

———. *Making History.* New York: Samuel French, 1989.

Fuccaro, Nelida. *Histories of City and State in the Persian Gulf: Manama since 1800.* Cambridge: Cambridge University Press, 2009.

Fung, Stanislaus. "Guide to Secondary Sources on Chinese Gardens." *Studies in the History of Gardens & Designed Landscapes* 18.3 (1998): 269–86.

Geertz, Clifford. *The Interpretation of Cultures: Selected Essays.* New York: Basic Books, 1973.

———. *Works and Lives: The Anthropologist as Author.* Stanford, CA: Stanford University Press, 1988.

Gerhardt, Christina. *Sea Change: An Atlas of Islands in a Rising Ocean.* Berkeley: University of California Press, 2023.

Ghirri, Luigi. *The Complete Essays, 1973–1991.* London: MACK, 2017.

Gobo, Giampietro, and Andrea Molle. *Doing Ethnography.* New York: Sage, 2017.

Goffman, Erving. "Regions and Region Behavior." In *The Presentation of Self in Everyday Life.* New York: Doubleday, 1990.

Goldman, Marcio. "How to Learn in an Afro-Brazilian Spirit Possession Religion: Ontology and Multiplicity in Candomblé." In *Learning Religion: Anthropological Approaches,* edited by Ramón Sarró and David Berliner, 103–19. Oxford: Berghahn Books, 2007.

Gordenstein, Samuel Lira. "Planting Axé in the City: Urban Terreiros and the Growth of Candomblé in Late Nineteenth-Century Salvador, Bahia, Brazil." *Journal of African Diaspora, Archaeology, and Heritage* 5 (2016): 71–101.

Grant, James. *The Critical Imagination.* Oxford: Oxford University Press, 2013.

Guattari, Félix. *The Three Ecologies* Translated by Ian Pindar and Paul Sutton. London: Athlone Press, 2000.

Gundaker, Grey, and Judith McWillie. *No Space Hidden: The Spirit of African American Yard Work.* Knoxville: University of Tennessee Press, 2005.

Harvey, David. *Justice, Nature, and the Geography of Difference.* Cambridge, MA: Blackwell, 1996.

Henry, Françoise. *Irish Art in the Early Christian Period, to 800 A.D.* Ithaca, NY: Cornell University Press, 1965.

Hester, Randolph T. *Design for Ecological Democracy.* Cambridge, MA: MIT Press, 2006.

Hilderbrand, Gary R. "Varied Tree Shade for New Urban Pleasures." *Harvard Design Magazine* 31 (Fall/Winter 200): 79–85.

Hockney, David. *David Hockney: A Bigger Picture.* Film by Bruno Wollheim. Coluga Pictures, 2017. https://vimeo.com/ondemand/.

Hockney, David and Martin Gayford, *A History of Pictures.* New York: Abrams, 2020.

Holes, Clive. *Colloquial Arabic of the Gulf and Saudi Arabia.* London: Routledge and Kegan Paul, 1984.

Holmes, Damian. "VPUU Harare Khyalitsha, Cape Town South Africa, Tarna Klitzner Landscape Architects." World Landscape Architecture (WLA) website, March 2014. https://worldlandscapearchitect.com/vpuu-harare-khyalitsha-cape-town-south-africa-tarna-klitzner-landscape-architects/.

Hunt, John Dixon. "Is Landscape History?" In *Is Landscape . . . ? Essays on the Identity of Landscape,* edited by Gareth Doherty and Charles Waldheim, 247–60. Abingdon, UK: Routledge, 2015.

Ingold, Tim. "Anthropology Is *Not* Ethnography." *Proceedings of the British Academy* 154 (2008): 69–92.

———. "Lines, Drawing, the Human Condition." A conversation between Tim Ingold, Momoyo Kaijima, Andreas Kalpakci, and Anh-linh Ngo. Translated from the German in *ARCH+* 238 (March 2020). *Drawing Matter website,* October 2021. https://drawingmatter.org/lines-drawings-the-human-condition/.

———. *Making: Anthropology, Archaeology, Art and Architecture.* Abingdon, UK: Routledge, 2013.

———. "The Temporality of the Landscape." *World Archaeology* 25.2 (1993): 152–74.

Ireland, Bunreacht na hÉireann. *Constitution of Ireland.* Dublin: Oifig an tSoláthair, 1937.

Jackson, Anthony. *Anthropology at Home.* London: Tavistock, 1987.

Jackson, Bruce. *Fieldwork.* Urbana: University of Illinois Press, 1987.

Jackson, John Brinckerhoff. *Landscape in Sight: Looking at America.* Edited by Helen Lefkowitz Horowitz. New Haven, CT: Yale University Press, 1997.

———. "The Necessity for Ruins." In *The Necessity for Ruins, and Other Topics,* 89–102. Amherst: University of Massachusetts Press, 1980.

———. "The Need of Being Versed in Country Things." *Landscape* 1.1 (Spring 1951): 1–5.

———. "The Word Itself." In *Discovering the Vernacular Landscape.* New Haven, CT: Yale University Press, 1986.

Jekyll, Francis, Edwin Landseer Lutyens, Agnes Jekyll, and Jonathan Cape. *Gertrude Jekyll: A Memoir.* London: Jonathan Cape, 1934.

Jekyll, Gertrude. *Children and Gardens.* London: Country Life, 1908.

———. *Colour Schemes for the Flower Garden.* 3rd ed. London: Country Life, 1914.

———. *Home and Garden: Notes and Thoughts, Practical and Critical, of a Worker in Both.* London: Longmans, Green, 1900.

———. *Wood and Garden: Notes and Thoughts, Practical and Critical, of a Working Amateur.* Godalming, UK: Longmans, Green, 1899; reprint Woodbridge, UK: Antique Collectors' Club, 1981.

Jellicoe, Geoffrey. *Gardens of the Mind. Television program,* directed and produced by Roger Last. Aired on February 17, 1991, on BBC Two.

———. *Geoffrey Jellicoe: The Studies of a Landscape Designer over 80 Years.* Vol. 1. Woodbridge, UK: Garden Art Press, 1993.

———. *The Guelph Lectures on Landscape Design.* Guelph, Ontario: University of Guelph, 1983.

———. "Ourselves and History." *Landscape Architecture Magazine 49.1* (Autumn 1958): 28–32.

———. "Ronald Tree and the Gardens of Ditchley Park: The Human Face of History." *Garden History* 10, no. 1 (Spring 1982): 80–91.

———. *Studies in Landscape Design. 3 vols.* London: Oxford University Press, vol. 1, 1960; vol. 2, 1966; vol. 3, 1970.

———. "A Table for Eight." In *Space for Living: Landscape Architecture and the Allied Arts and Professions,* eited by Sylvia Crowe, 13–21. Amsterdam: Djambatan, 1961.

Jellicoe, Geoffrey, and Susan Jellicoe. *Landscape of Man: Shaping the Environment from Prehistory to the Present Day.* London: Thames and Hudson, 1975.

Job, David. *New Directions in Geographical Fieldwork.* Cambridge: Cambridge University Press, 1999.

John-Alder, Kathleen. *Ian McHarg and the Search for Ideal Order.* Abingdon, UK: Routledge, 2020.

Khuri, Fuad. *An Invitation to Laughter: A Lebanese Anthropologist in the Arab World.* Edited by Sonia Jalbout Khuri. Chicago: University of Chicago Press, 2007.

Kirkwood, Niall G. *The Art of Landscape Detail: Fundamentals, Practices, and Case Studies.* New York: Wiley, 1999.

———. "Curating Resources." In *Ecological Urbanism,* edited by Mohsen Mostafavi and Gareth Doherty, 196–99. Zürich: Lars Müller, 2016.

———. *Weathering and Durability in Landscape Architecture: Fundamentals, Practices, and Case Studies.* Hoboken, NJ: John Wiley, 2004.

Lakoff, George, and Mark Johnson. *Metaphors We Live By.* Chicago, IL: University of Chicago Press, 1980.

Landes, Ruth. *City of Women.* Albuquerque: University of New Mexico Press, 1994.

Landscape Architectural Accreditation Board, *Accreditation Standards for Professional Programs in Landscape Architecture.* Washington, DC: American Society of Landscape Architects, 2021.

Larsen, Curtis E. *Life and Land Use on the Bahrain Islands: The Geoarchaeology of an Ancient Society.* Chicago: University of Chicago Press, 1983.

Latour, Bruno. "Forty Years Later—Back to a Sub-lunar Earth." In *Ecological Urbanism, rev. ed., edited by* Mohsen Mostafavi and Gareth Doherty, 130–33. Zürich: Lars Müller, 2016.

Lino e Silva, Moisés. *Minoritarian Liberalism: A Travesti Life in a Brazilian Favela.* Chicago: University of Chicago Press, 2022.

Lucas, A. T. *Furze: A Survey and History of Its Uses in Ireland.* Dublin: National Museum of Ireland, 1960.

Lynch, Kevin, and Gary Hack. *Site Planning.* 3rd ed. Cambridge, MA: MIT Press, 1984.

Machado, Rodolfo. *The Favela-Bairro Project: Jorge Mario Jáuregui Architects.* Cambridge, MA: Harvard University Graduate School of Design, 2008.

Maghtochair (pseud. of Michael Harkin). *Inishowen: Its History, Traditions, & Antiquities Containing a Number of Original Documents, with Numerous Notes from the Annals of the Four Masters, and Other Sources.* Londonderry: [Derry] Journal Office, 1867.

Makhzoumi, Jala. "Beirut's Public Realm and the Discourse of Landscape Citizenships." In *Landscape Citizenships,* edited by Tim Waterman, Jane Wolff, and Ed Wall, 182–204. London: Routledge, 2021.

———. "Colonizing Mountain, Paving Sea: Neoliberal Politics and the Right to Landscape in Lebanon." In *The Right to Landscape: Contesting Landscape and Human Rights,* edited by Shelley Egoz, Jala Makhzoumi, and Gloria Pungetti, 227–44. Florence: Taylor & Francis, 2011.

———. "Re-Defining Landscape: Erbil's Urban-Rural Greenbelt." *Perspecta: The Yale Architectural Journal* 53 (2020): 14–31.

———. "Reflections on Landscape and Landscape Architecture Education in the Arab Middle East." In *The Routledge Handbook of Landscape Architecture Education,* edited by Diedrich Bruns and Stefanie Hennecke, 303–14. New York: Routledge, 2022.

Manfredi, Victor. "*Ẹ̀là kò là!* The Disneyfication of Ọ̀ṣun Òṣogbo." Website of African Studies Center, Boston University. http://people.bu.edu/manfredi/JesuOs.ogboEvirato.pdf.

Marcus, George E. "Ethnography in/of the World System: The Emergence of Multi-Sited Ethnography." *Annual Review of Anthropology* 24.1 (1995): 95–117.

Mathur, Anuradha, and Dilip da Cunha. *Mississippi Floods: Designing a Shifting Landscape.* New Haven, CT: Yale University Press, 2001.

Matory, J. Lorand. *Black Atlantic Religion: Tradition, Transnationalism, and Matriarchy in the Afro-Brazilian Candomblé.* Princeton, NJ: Princeton University Press, 2005.

McEvoy-Levy, Siobhan. "Youth Spaces in Haunted Places: Placemaking for Peacebuilding in Theory and Practice." *International Journal of Peace Studies* 17.2 (2012): 1–32.

McHarg, Ian L. *Design with Nature.* 1st ed. Garden City, NY: Natural History Press, for American Museum of Natural History, 1969.

———. "Human Ecological Planning at Pennsylvania." *Landscape Planning* 8 (1981): 109–20.

Menon, Dilip M. *Changing Theory: Concepts from the Global South.* Milton Park, UK: Routledge, 2022.

Mehrotra, Rahul. *Architecture in India: Since 1990.* Mumbai: Hatje Cantz, 2011.

Merrill, Ross. "Advice on Plein Air Equipment." *American Artist* 73.802 (2009): 28–30.

Meyer, Elizabeth. "Site Citations: The Grounds of Modern Landscape Architecture." In *Site Matters: Design Concepts, Histories, and Strategies, edited by* C. Burns and A. Kahn, 95–129. New York: Taylor & Francis, 2005; reprint 2021.

Mills, Albert J., Gabrielle Durepos, and Elden Wiebe. *Encyclopedia of Case Study Research.* Los Angeles, CA: Sage, 2010.

Moggridge, Hal. "Jellicoe, Sir Geoffrey Alan (1900–1996), Landscape Architect." *Oxford Dictionary of National Biography.* Website of Oxford Dictionary of National Biography, September 23, 2004, rev. May 26, 2005. https://doi.org/10.1093/ref:odnb/40519.

Momoyo Kaijima, Laurent Stalder, and Yu Iseki, eds. *Architectural Ethnography.* Tokyo: Toro Kato, 2018.

Monaghan, John, and Peter Just. *Social and Cultural Anthropology: A Very Short Introduction.* Oxford: Oxford University Press, 2000.

Mostafavi, Mohsen, and Gareth Doherty, eds. *Ecological Urbanism.* Zürich: Lars Müller, 2010; rev. ed. 2016.

Mostafavi, Moshen, and Ciro Najle, eds. *Landscape Urbanism: A Manual for the Machinic Landscape.* London: AA Publications, 2003.

Murphy, Keith M., and George E. Marcus. "Epilogue: Ethnography and Design, Ethnography in Design . . . Ethnography by Design." In *Design Anthropology: Theory and Practice,* edited by Wendy Gunn, Ton Otto, and Rachel Charlotte Smith, 251–68. London: Bloomsbury, 2013.

Murrell, Nathaniel Samuel. "Obeah: Magical Art of Resistance." In *Afro-Caribbean Religions: An Introduction to Their Historical, Cultural, and Sacred Traditions.* Philadelphia, PA: Temple University Press, 2010.

Nadin, Mihai. *Anticipation: The End Is Where We Start From.* Baden, Switzerland: Lars Müller, 2002.

Ndubisi, Forster O. "Introduction." In *The Ecological Design and Planning Reader,* edited by Forster O. Ndubisi, 1–10. Washington, DC: Island Press, 2014.

Odbert, Chelina, and Joseph Mulligan. "The Kibera Public Space Project: Participation, Integration, and Networked Change." In *Now Urbanism: The Future City Is Here,* edited by Jeffrey Hou, Benjamin Spencer, Thaïsa Way, and Ken Yocom, 177–92. Abingdon, UK: Routledge, 2015.

Ogunfolakan, A., C. Nwokeocha, A. Olayemi, M. Olayiwola, A. Bamigboye, A. Olayungbo,

Ogiogwa, O. Oyelade, and O. Oyebanjo, O. "Rapid Ecological and Environmental Assessment of Osun Sacred Forest Grove, Southwestern Nigeria." *Open Journal of Forestry* 6 (2016): 243–258. doi: 10.4236/ojf.2016.64020.

Olin, Laurie. *Across the Open Field: Essays Drawn from English Landscapes.* Philadelphia: University of Pennsylvania Press, 2000.

Oliveira, Rafael Soares de. "Redes e Territórios: uma investigação hermenêutica sobre a história do Candomblé da Berroquinha." *Culture Vozes* 4.94 (2001): 44–61.

Olmos, Margarite Fernandez. *Creole Religions of the Caribbean: An Introduction from Vodou and Santeria to Obeah and Espiritismo.* New York: New York University Press, 2011.

Olmsted, Frederick Law. "The People's Park at Birkenhead (1851)." In *Frederick Law Olmsted: Essential Texts,* edited by Robert Twombly, 39–48. New York; London: W. W. Norton, 2010.

———. "The Phalanstery and the Phalansterians (1852)." In *Frederick Law Olmsted: Essential Texts,* edited by Robert Twombly, 49–57. New York; London: W. W. Norton, 2010.

Olmsted, Frederick Law Sr., and Calvert Vaux. "Description of a Plan for the Improvement of the Central Park, 'Greensward,' 1858." In *Forty Years of Landscape Architecture: Central Park,* edited by Frederick Law Olmsted, Jr and Theodora Kimball, 214–33. Cambridge, MA: MIT Press, 1973.

Olupona, Jacob K. *City of 201 Gods: Ilé-Ifè in Time, Space, and the Imagination.* Berkeley: University of California Press, 2011.

Oxóssi, Mãe Stella de (writing as Maria Stella de Azevedo Santos). *Meu tempo é agora.* Salvador da Bahia: Assembleia Legislativa do Estado da Bahia, 2010.

———. *O que as folhas cantam (For whom the leaves sing).* Salvador: Ossos do Ofício, 2014.

Pamuk, Orhan. *My Name Is Red.* Translated from the Turkish by Erdag M. Göknar. New York: Faber & Faber, 2001.

Pandian, Anand. *A Possible Anthropology: Methods for Uneasy Times.* Durham, NC: Duke University Press, 2019.

Pavlidis, Stephen J. *The Exuma Guide: A Cruising Guide to the Exuma Cays.* Port Washington, WI: Seaworthy Publications, 2010.

Peirano, Mariza G. S. "When Anthropology Is at Home: The Different Contexts of a Single Discipline." *Annual Review of Anthropology* 27.1 (1998): 105–28.

Pessoa de Barros, José Flávio, and Eduardo Napoleão. *Ewé Òrìsà: Uso litúrgico e terapêutico dos vegetais nas casas de Candomblé Jêje-Nagô.* Rio de Janeiro: Editora Bertrand Brasil, 1999.

Racin, Liat, and Gareth Doherty. "Cultivating Fruit: An Account of Applied Field Research in the Exuma Archipelago, The Bahamas." *Landscape Architecture Frontiers* 5.2 (2017): 86–101.

Ragsdale, Joseph. "Fieldwork Hybrids: Learning from Other Disciplines How to Read, Record, and Reveal the Landscape." *Landscape Research Record* 7. Blacksburg, VA: CELA Conference, 2018.

Read, Herbert. *A Concise History of Modern Painting.* New York: Praeger, 1959.

Reed, Christopher. "Nearsighted and Visionary." Review of *Gertrude Jekyll* by Sally Festing (New York: Viking, 1991), *New York Times,* May 24, 1992, sec. 7, p. 20,

Rego, Jussara. "Territórios do candomblé: a desterritorialização dos terreiros na Região Metropilitana de Salvador, Bahia," *GeoTextos* 2.2 (2006): 31–85.

Rossi, Aldo. *A Scientific Autobiography.* Cambridge, MA: MIT Press, 1981.

Rybczynski, Witold. *A Clearing in the Distance: Frederick Law Olmsted and America in the Nineteenth Century.* New York: Scribner, 1999.

———. *Last Harvest: How a Cornfield Became New Daleville: Real Estate Development in America from George Washington to the Builders of the Twenty-First Century, and Why We Live in Houses Anyway.* New York: Scribner, 2007.

Sanjek, Roger. "A Vocabulary for Fieldnotes." In *Fieldnotes—The Makings of Anthropology,* edited by Roger Sanjek, 92–121 Ithaca, NY: Cornell University Press, 1990.

Savishinsky, Joel E. *Strangers No More: Anthropological Studies of Cat Island, The Bahamas.* Ithaca, NY: Ithaca College, Department of Anthropology, 1978.

Sayre, Roger. *Nature in Focus: Rapid Ecological Assessment.* Washington, DC: Island Press, 2000.

Schmidt, Bettina E. *Spirits and Trance in Brazil: An Anthropology of Religious Experience.* London: Bloomsbury, 2017.

Sealey, Neil E. *The Bahamas Today: An Introduction to the Human and Economic Geography of the Bahamas.* London: Macmillan Caribbean, 1990.

———. *Bahamian Landscape: Introduction to the Geology and Physical Geography of the Bahamas.* 3rd ed. New York: Macmillan, 2006.

Sheller, Mimi. *Citizenship from Below: Erotic Agency and Caribbean Freedom.* Durham, NC: Duke University Press, 2012.

Shepherd, J. C., and G. A. Jellicoe. *Italian Gardens of the Renaissance.* 3rd ed. 1925; London: Academy Editions, 1986.

Sheridan, Michael J., and Celia Nyamweru, eds. *African Sacred Groves: Ecological Dynamics and Social Change.* Athens: Ohio University Press, 2007.

Simon, Herbert. *The Sciences of the Artificial.* Cambridge, MA: MIT Press, 1969.

Stanford University and Center for the Study of Language Information. *Stanford Encyclopedia of Philosophy.* Stanford, CA: Stanford University, 1997: https://plato.stanford.edu.

Sousa Júnior, Vilson Caetano de. *Corujebó: candomblé e polícia de costumes, 1938–1976.* Salvador da Bahia: EDUFBA, 2018.

———. "*Kosi Ewé, Kosi Orisa:* The Use of Plants in Afro-Brazilian Religions." Presentation to the Sacred Groves and Secret Parks: Orisha Landscapes in Brazil and West Africa colloquium, Graduate School of Design, Harvard University, October 3–4, 2019, https://www.gsd.harvard.edu/event/sacred-groves-secret-parks-orisha-landscapes-in-brazil-and-west-africa/.

Spens, Michael. "Admirable Jellicoe." *Architectural Review* 186.1111 (1989): 85–92.

———. *The Complete Landscape Designs and Gardens of Geoffrey Jellicoe.* London: Thames and Hudson, 1994.

———. *Gardens of the Mind: The Genius of Geoffrey Jellicoe.* Woodbridge, UK: Garden Art Press, 1992.

———. *Jellicoe at Shute.* London: Academy Editions, 1993.

Spirn, Anne Whiston. *The Eye Is a Door: Landscape, Photography, and the Art of Discovery.* Nahant, MA: Wolf Tree Press, 2014. Kindle.

———. *The Granite Garden: Urban Nature and Human Design.* New York: Basic Books, 1984.

———. *The Language of Landscape.* New Haven, CT: Yale University Press, 1998.

Steiner, Frederick R. *Human Ecology: How Nature and Culture Shape Our World.* 2nd ed. Washington, DC: Island Press, 2016.

———. *Making Plans: How to Engage with Landscape, Design, and the Urban Environment.* Austin: University of Texas Press, 2018.

Steinitz, Carl. "From Project to Global: On Planning and Scale." *Landscape Research* 9.2 (2004): 117–27.

Stevens, Quentin. *The Ludic City.* Abingdon, UK: Routledge, 2007.

Stilgoe, John. "Design Standards: Whose Meanings?" In *Regulating Place: Standards and the Shaping of Urban America,* edited by Eran Ben-Joseph and Terry S. Szold, 17–44. New York: Routledge, 2005.

———. Introduction to *Lifeboat: A History of Courage, Cleverness, and Survival at Sea,* by John Stilgoe, 1–19. Charlottesville: University of Virginia Press, 2003.

———. *Outside Lies Magic: Regaining History and Awareness in Everyday Places.* New York: Walker, 1998.

Strathern, Marilyn. *Commons and Borderlands: Working Papers on Interdisciplinarity, Accountability and the Flow of Knowledge.* Wantage, UK: Sean Kingston, 2004.

Strongman, Roberto, *Queering Black Atlantic Religions: Transcorporeality in Candomblé, Santería, and Vodou.* Durham, NC: Duke University Press, 2019.

Tan, Ekim. "Negotiation and Design for the Self-Organizing City: Gaming as a Method for Urban Design." PhD diss., Delft University of Technology, 2014.

———. *Play The City – Games Informing the Urban Development.* Prinsenbeek, Netherlands: Jap San Books, 2017.

Tankard, Judith. "Where Flowers Bloom in the Sands." *Country Life,* March 12, 1998, 82–85.

Textor, Robert B. Introduction to *The World Ahead: An Anthropologist Anticipates the Future, by* Margaret Mead, edited and annotated by Robert B. Textor, 1–31. New York: Berghahn Books, 2005.

Treib, Marc. "Kritische Haltung Zeigen — Being Critical," *Topos* 49 (2004), 8.

———. "Landscape Architecture Criticism." *Land Forum,* edited by Anita Berrizbeitia. (Fall/Winter 1997): 9–10.

Tuan, Yi Fu. Foreword to *Geography and the Human Spirit,* by Anne Buttimer. Baltimore, MD: Johns Hopkins Press, 1993.

Turner, Victor W. *The Forest of Symbols: Aspects of Ndembu Ritual.* Ithaca, NY: Cornell University Press, 1967.

Ur, Jason. "CORONA Satellite Photography and Ancient Road Networks: A Northern Mesopotamian Case Study." *Antiquity* 77.295 (2003): 102–15.

van de Port, Mattijs. "Candomblé in Pink, Green and Black: Re-scripting the Afro-Brazilian Religious Heritage in the Public Sphere of Salvador, Bahia." *Social Anthropology* 13.1 (2005), 3–26.

———. *Ecstatic Encounters: Bahian Candomblé and the Quest for the Really Real.* Amsterdam: University of Amsterdam Press, 2011.

Velame, Fábio Macedo. "Arquiteturas dos Terreiros de Candomblé: Entre os Africanistas e Crioulistas." *Anais do Seminário Arquitetura Vernácula/Popular. Anais.* Salvador (BA): Programa de Pós-Graduação em Arquitetura e Urbanismo da UFBA, 2021.

Venturi, Robert, Denise Scott Brown, and Steven Izenour, *Learning from Las Vegas: The Forgotten Symbolism of Architectural Form.* Cambridge, MA: MIT Press, 1977.

Verger, Pierre Fatumbi. *Ewé, the Use of Plants in Yoruba Society.* São Paulo: Editora Schwarcz, 1995.

———. *Notas sobre o culto aos orixás e voduns na Bahia de Todos os Santos, no Brasil, e na antiga costa dos escravos, na África.* São Paulo: EDUSP, 1999.

Vogt, Günther. *Landscape as a Cabinet of Curiosities: In Search of a Position.* Edited by Rebecca Bornhauser and Thomas Kissling. Zürich: Lars Müller, 2015.

Wafer, Jim. *The Taste of Blood: Spirit Possession in Brazilian Candomblé.* Philadelphia: University of Pennsylvania Press, 1992.

Waldheim, Charles, "Aerial Representation and the Recovery of Landscape." In *Recovering Landscape,* edited by James Corner, 121–40. New York: Princeton Architectural Press, 1999.

———. *CASE: Hilberseimer/Mies van der Rohe, Lafayette Park Detroit.* Munich: Prestel; Cambridge, MA: Harvard University Graduate School of Design, 2004.

Wall, Ed. "Incompleteness: Landscapes, Cartographies, Citizenships." *Landscape Research* 47, no. 2 (2022): 179–94.

———. "Working with Uncertainties: Living with Masterplanning at Elephant and Castle." In *Landscape Citizenships,* edited by Tim Waterman, Jane Wolff, and Ed Wall, 205–24. London; New York: Routledge, 2021.

Waly, Tarek. *Private Skies: The Courtyard Pattern in the Architecture of the House.* Bahrain: Al Handasah Center, 1992.

Waterman, Tim. "Making Meaning: Utopian Method for Minds, Bodies, and Media in Architectural Design." *Open Library of Humanities* 4, no. 1 (2018): 1–26.

Wenger, Suzanne. *The Timeless Mind of the Sacred: Its New Manifestation in the Osun Groves.* Ibadan, Nigeria: University of Ibadan, 1977.

Wenger, Suzanne, and Gert Chesi. *A Life with the Gods: In Their Yoruba Homeland.* Innsbruck, Austria: Perlinger, 1983.

Whyte, William Hollingsworth. *The Social Life of Small Urban Spaces.* Washington, DC: Conservation Foundation, 1980.

Widdowson, John. "Fieldwork." In *The Handbook of Secondary Geography,* edited by M. Jones, 228–41. Sheffield, UK: Geographical Association, 2017.

Wilde, Oscar. *The Critic as Artist.* New York: David Zwirner Books, 2019.

Wilson, Thomas M., and Hastings Donnan. "Nation, State and Identity at International Borders." In *Border Identities: Nation and State at International Frontiers, edited by Thomas M. Wilson and Hastings Donnan,* 1–30. Cambridge: Cambridge University Press, 1998.

Wolcott, H. F. *The Art of Fieldwork.* Walnut Creek, CA: AltaMira Press, 2005.

Wood, Denis. *The Power of Maps.* New York: Guilford Press, 1992.

Woudstra, Jan, and Colin Roth. *A History of Groves.* Abingdon, UK: Taylor and Francis, 2017.

Yamin, R., and K. Bescherer Metheny. *Landscape Archaeology: Reading and Interpreting the American Historical Landscape.* Knoxville: University of Tennessee Press, 1996.

Yaneva, Albena. "Architectural Theory at Two Speeds." Website of Ardeth. October 1, 2017, http://journals.openedition.org/ardeth/983.

———. *Crafting History: Archiving and the Quest for Architectural Legacy.* Ithaca, NY: Cornell University Press, 2020.

———. "How to Study the Ecology of Practice." In *Five Ways to Make Architecture Political: An Introduction to the Politics of Design Practice.* London: Bloomsbury, 2018.

———. *Made by the Office for Metropolitan Architecture: An Ethnography of Design.* Rotterdam: 010 Uitgeverij, 2009.

Yang, Bo. *Landscape Performance.* Abingdon, UK: Taylor and Francis, 2018.

Yentsch, Anne Elizabeth. "Introduction: Close Attention to Place—Landscape Studies by Historical Archaeologists." In *Landscape Archaeology: Reading ad Interpreting the American Historical Landscape,* edited by Rebecca Yamin, xxiii–xlii. Knoxville: University of Tennessee Press, 1996.

Yin, Robert K., *Case Study Research and Applications: Design and Methods.* 6th ed. Thousand Oaks, CA: Sage, 2018.

INDEX

Italicized page numbers refer to figures.